# EPA 608

## STUDY GUIDE

The Complete Blueprint to Certification Success:
Comprehensive Test Prep, In-Depth Review, Expert Insights, and Practice Questions for Achieving EPA 608 Certification

**2024-25**

- ✓ TOPIC FLOW STRICTLY AS PER THE TEST SYLLABUS (150+ PAGES)
- ✓ EXAM PATTERN ANALYSIS
- ✓ SCORE-MAXIMIZING STRATEGIES
- ✓ PRACTICE TEST CORE, TYPE 1, TYPE 2 AND TYPE 3
- ✓ GLOSSARY & 20+ TABLES

**DR. FANATOMY**

# copyright@ dr. fanatomy 2023

All rights reserved. No part of this publication may be reproduced, distributed, or transmitted in any form or by any means, including photocopying, recording, or other electronic or mechanical methods, without the prior written permission of the publisher, except in the case of brief quotations embodied in critical reviews and certain other noncommercial uses permitted by copyright law.

This book is a work of non-fiction, and any resemblance to actual persons, living or dead, or actual events is purely coincidental.

The information and techniques described in this book are intended for educational and informational purposes only. The author and publisher shall not be held liable for any injury, damage, or loss arising from using or misusing the information presented in this book.

While every effort has been made to ensure the accuracy of the information contained within this book, the author and publisher make no warranties or representations express or implied, about the completeness, accuracy, reliability, suitability, or availability with respect to the contents of this book for any purpose. The use of any information provided in this book is at the reader's own risk.

# Declaration

This book is an independent work and has **no official affiliation with the Environmental Protection Agency (EPA)** or any related authorities. This study guide is not an official version nor endorsed or published by the EPA.

The primary purpose of this book is to provide valuable insights, helpful tips, and effective study techniques based on our experience and knowledge in the HVAC-R industry. We have strived to create a comprehensive resource that aids aspiring professionals in preparing for the EPA 608 certification exam.

It is essential to understand that while this guide is designed to assist you in your exam preparation, it still needs to **replace the official study materials** provided by the EPA. We strongly advise candidates to refer to the EPA's official documents and guidelines for complete and up-to-date information on the certification exam.

As an **unofficial guide**, this book is intended to supplement formal training or professional education. The EPA 608 certification is a significant milestone in the HVAC-R field, and successful candidates must also demonstrate practical skills and adhere to ethical standards in their profession.

Although we have made every effort to ensure the accuracy of the information presented in this book, the author and publisher cannot guarantee its completeness, accuracy, reliability, suitability, or availability for any purpose. Any information from this book is solely at the reader's discretion and risk.

In conclusion, the "EPA 608 Study Guide" is an independent effort to support your exam preparation journey. We wish you the best of luck and success in achieving your certification goals.

# TABLE OF CONTENTS

## 1. ALL ABOUT THE EXAM- INTRODUCTION (4-12)

- What is EPA 608 Certification
- Refrigerant Circuit's Integrity
- Types of EPA 608 Certification
- Who is a Technician as per EPA regulations?
- Types of Technicians
- Exam Pattern-EPA 608 Certification
- Additional Information: EPA 608 Exam

## 2. CORE EXAM OF EPA 608 CERTIFICATION (13-64)

- Scope of Core Section
- Exam-specific points for the Core Section
- Environmental Impacts
- Clean Air Act and Montreal Protocol
- Section 608 Regulations
- Substitute Refrigerants and oils
- Refrigeration
- Three R Definitions
- Recovery Techniques
- Dehydration Evacuation
- Safety
- Shipping

## 3. TYPE 1 (SMALL APPLIANCES) OF EPA 608 CERTIFICATION (65-77)

- Scope of Type I Section
- Recovery Requirements
- Recovery techniques
- Safety

## 4. TYPE 2 (HIGH-PRESSURE) OF EPA 608 CERTIFICATION (78-93)

- Scope of Type II Section
- Leak Detection
- Leak repair requirements
- Recovery Techniques
- Recovery requirements
- Refrigeration
- Safety

## 5. TYPE 3 (LOW-PRESSURE) OF EPA 608 CERTIFICATION (94-110)

- Scope of Type III Section
- Leak Detection
- Leak repair requirements
- Recovery Techniques
- Recharging Techniques
- Recovery Requirements

- Refrigeration
- Safety

## 6. PRACTICE TESTS (111-146)

- Core Section
- Type 1
- Type 2
- Type 3

## 7. GLOSSARY & CONCLUSION (147-150)

- Glossary
- Conclusion

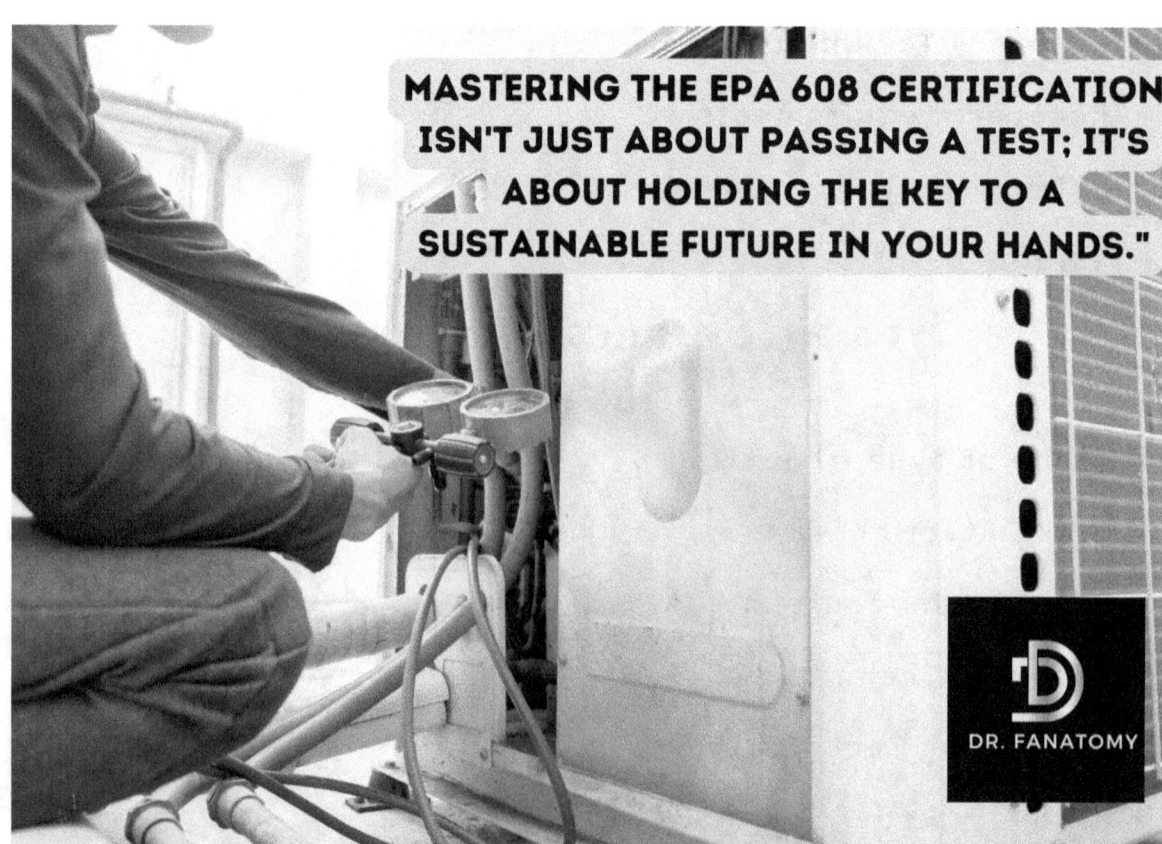

"MASTERING THE EPA 608 CERTIFICATION ISN'T JUST ABOUT PASSING A TEST; IT'S ABOUT HOLDING THE KEY TO A SUSTAINABLE FUTURE IN YOUR HANDS."

# 1. Introduction to EPA 608 Certification

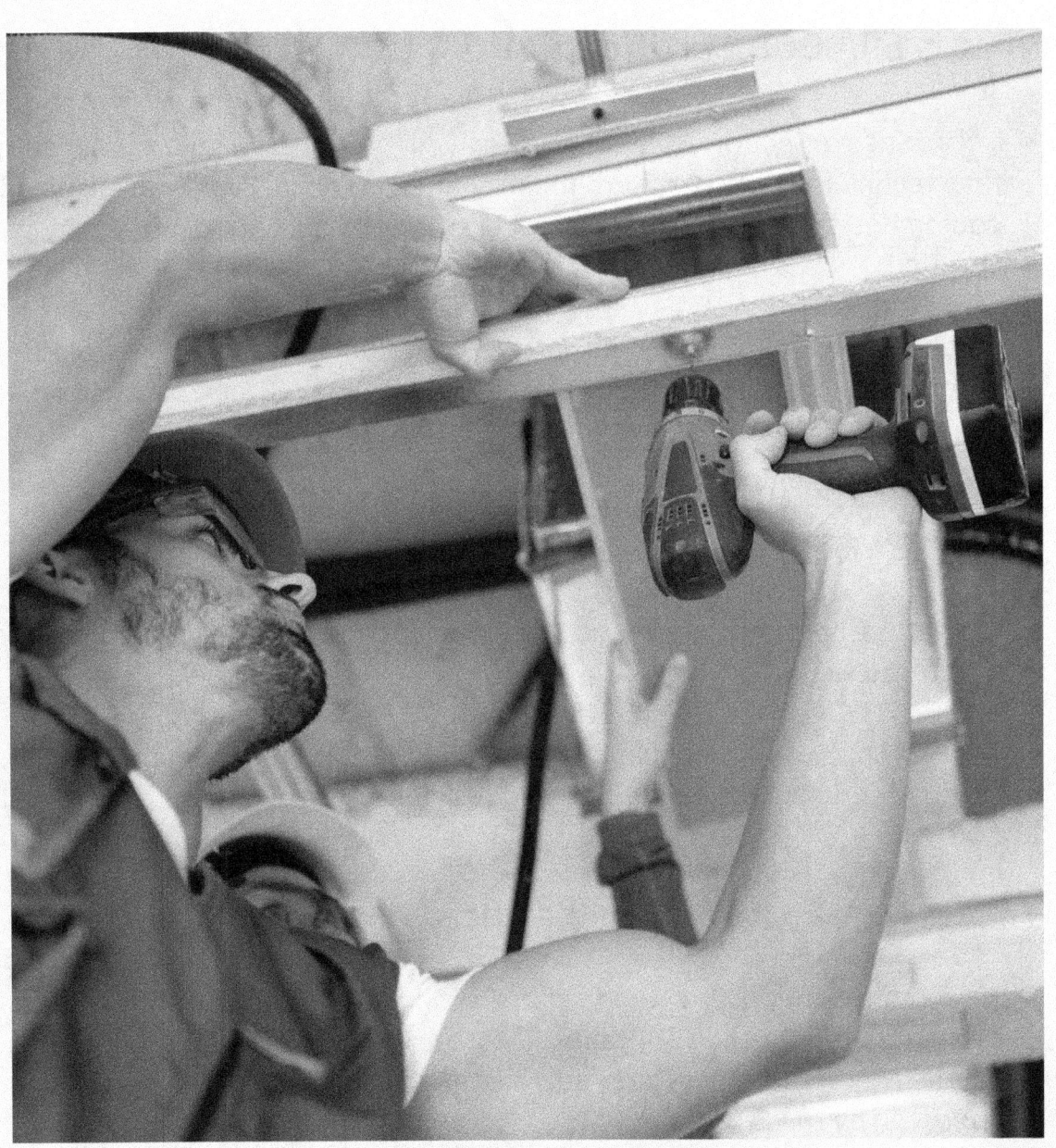

# 1. EPA 608 CERTIFICATION: AN INTRODUCTION

This chapter focuses on inquiring about all the necessary preliminary information from a potential candidate.

## What is EPA 608 Certification

- According to Section 608 of the Clean Air Act, EPA regulations mandate that **technicians** who perform **maintenance, service, repair, or disposal** of equipment that may release refrigerants into the atmosphere must be certified.

- To obtain Section 608 Technician Certification, **technicians** must pass an EPA-approved test specific to the equipment they intend to work on. It is not permitted to work under someone else's certification.

- An EPA-approved certifying organization must administer the test. EPA 608 Technician Certification credentials never expire.

- To get a Universal Certification, technicians must take the **core test** as a **proctored exam.**

- Those who deal with appliances containing **regulated refrigerants** must be certified in proper handling techniques under the Clean Air Act's Section 608.

- Regulated Refrigerants include **CFC, HCFC, HFC,** and **HFO** refrigerants.

- To acquire Section 608 Technician Certification, technicians must pass a test approved by the EPA **tailored to the type of appliance** they plan to handle. The test must be administered by a certifying organization approved by the EPA.

# Refrigerant Circuit's Integrity

Activities that could damage the refrigerant circuit's integrity and may release refrigerants into the atmosphere

- *Attaching and detaching hoses and gauges to add or remove refrigerant or measure pressure.*

- *Adding or removing refrigerant from the appliance and replacing components of the refrigeration circuit, such as the compressor, condenser, evaporator, expansion (throttling) device, or filter drier.*

Activities that can be done without harming the refrigerant circuit of an appliance:

- *Painting the appliance*

- *Cleaning the exterior of the evaporator or condenser coils*

- *Replacing fans or blowers*

- *Straightening heat exchanger fins*

- *Rewiring an electrical circuit*

- *Replacing insulation on a length of pipe*

- *Replacing a faulty capacitor, contactor, or potential relay*

- *Tightening nuts and bolts on the appliance*

- *You wouldn't typically expect to release refrigerants if you perform maintenance, service, repair, or disposal on a device with refrigerant already removed. The only exception would be if the work involves adding or removing refrigerant from the appliance.*

## Types of EPA 608 Certification

| Certification Type | For Servicing | Exam Paper (Each paper has 25 questions and 70% passing marks) |
|---|---|---|
| Type I | small appliances | Core + Type I |
| Type II | high- or very high-pressure appliances, except small appliances and MVACs | Core + Type II |
| Type III | low-pressure appliances | Core + Type III |
| Universal | all types of equipment | Core + Type I+ Type II + Type III |

## Who is a Technician as per EPA regulations?

- EPA regulation for the Clean Air Act's Section 608 defines a technician as someone who does tasks like connecting hoses and gauges to measure pressure in an appliance, adding or removing refrigerant from an appliance, or doing anything that could harm a motor vehicle air conditioner or small appliance.

- Apprentices don't need certification if they work under a certified technician's supervision.

# Types of Technicians :

### Type I Technicians :

- Individuals who work with small appliances, such as water coolers, window units, refrigerators, freezers, dehumidifiers, residential ice machines, and package terminal air conditioning, must be certified as Type I technicians.

- A small appliance is a pre-assembled unit that is hermetically sealed and factory charged with 5 lbs. or less refrigerant.

- It is important to note that split systems are not included in Type I certification.

### Type 2 Technicians :

- Individuals certified as Type II technicians are responsible for maintaining, servicing, or repairing appliances that operate at medium, high, or very high pressures.

- This certification is required for those who dispose of such appliances and those who work on systems that do not fall under small devices or low-pressure systems (including motor vehicle air conditioners). To become certified, technicians must pass the Type II and Core exams.

- Those who receive passing scores are authorized to recover refrigerant while performing maintenance, service, or repair on medium- to high-pressure equipment and very high-pressure refrigerants, such as CFC-13 and CFC-503.

### Type 3 Technicians :

- Individuals responsible for maintaining, servicing, or repairing low-pressure appliances, or disposing of such appliances that may potentially release refrigerants into the atmosphere, must possess the appropriate certification as either Type III or Universal technicians.

**Universal Technician**

- Having Type I, II, and III certifications grants you a Universal certification card for servicing all appliances covered by those certifications.

- However, this card does not cover MVAC motor vehicle air conditioning certification. You need a separate Section 609 MVAC certification for that.

## Exam Pattern-EPA 608 Certification

**For all papers : (Core, Type I, Type II & Type III)**

- 25 questions in each paper

- Passing Marks (70%): 18/25 correct for every paper

- No Negative Marking

- Passing this section is compulsory for acquiring any certification of EPA 608 ( Type I, II, III, or Universal)

**(1) Core Section Exam : (25 Questions)**

- The Core has 25 questions covering ozone depletion, the Clean Air Act, the Montreal Protocol, refrigerant recovery, safety, and shipping.

- If you take the Core exam in an open-book format and later wish to obtain a Type II, Type III, or Universal certification, you must retake the 25 Core exam questions under proctored conditions.

**(2) Type I Exam : (25 Questions)**

- Type I exam consists of 25 sector-specific questions related to small appliances.

(3) Type II Exam : (25 Questions)

- Type II contains 25 sector-specific questions about medium and high-pressure appliances.

3) Type III Exam : (25 Questions)

- Section III comprises of 25 questions that are specific to low-pressure appliances.

| # | Core Exam | Type I Exam | Type II Exam | Type III Exam |
|---|---|---|---|---|
| Type I Certification | Pass (At least 18/25) | Pass (At least 18/25) | Not Applicable | Not Applicable |
| Type II Certification | Pass (At least 18/25) | Not Applicable | Pass (At least 18/25) | Not Applicable |
| Type III Certification | Pass (At least 18/25) | Not Applicable | Not Applicable | Pass (At least 18/25) |
| Universal | Pass (At least 18/25) | Pass (At least 18/25) | Pass (At least 18/25) | Pass (At least 18/25) |

# Additional Information: EPA 608 Exam

- According to federal regulations, an authorized test administrator (Proctor) must conduct this exam as a closed-book exam.

- A pressure-temperature (PT) chart and a calculator are the only external materials permitted during the test. Exams are closed-book tests.

- During the examination, it is strictly prohibited to use phones, and they must be turned off and put away. Any other electronic communication device should also not be used, and attempting to copy, distribute, or share photos of exam questions may result in revocation of certification and will be reported.

The exam requires certain personal information from technicians. Therefore, they should be ready to provide:

- *ID Card with photograph (Proctors will ask for this to verify your identity– this is required.)*

- *SSN (Social security number )*

- *Home/mailing address*

- *Date of Birth*

- *Phone number*

- *Email address*

- All individuals taking the exam will be listed in an online registry with their name, city, state, and certification attained. However, no personal information will be publicly displayed in the registry.

- A technician can take Core and any Type I, II, or III exams. It's not mandatory to take all four sections of the test.

- Before starting the exam, you must provide personal details and generate a distinct identification number. This number will be displayed in Social Security format on the front of your certification card. It is important to note that you should avoid using your Social Security number while filling out this section.

**Tips**

- Understand the question by paying attention to important words like "always," "never," "minimum," and "maximum."

- Examine all answer choices to maintain accuracy. Consider all possible answers to avoid missing questions. You can skip and return to questions if needed.

- Taking practice tests to prepare for your upcoming exam is a good idea. While they may not include the exact questions you'll encounter on the test, they can still help you prepare for a multiple-choice format and give you an idea of the topics that may be covered.

- It is advisable to read the entire question before choosing an answer. This will help you grasp the question better and make a more informed decision when selecting from the answer choices.

- It is essential to carefully review all available options before making a selection. Remember that the best answer may only sometimes be the first one presented. Take your time to fully consider each option before making a decision.

- Use the process of elimination to narrow down answer choices. Stick with the initial choice for the best results.

- Answer the easy questions first, then focus on the harder ones later.

# 2. Core Exam of EPA 608 Certification

# 2. CORE SECTION OF EPA 608 CERTIFICATION

This chapter focuses on the scope of the Core section and its explanation.

## Scope of Core Section (Syllabus for the Core Exam)
(source: https://www.epa.gov/section608)

### (1) Environmental Impacts

- How chlorine impacts the ozone layer

- Chlorine in chlorofluorocarbon (CFC) and hydrochlorofluorocarbon (HCFC) refrigerants

- Identifying CFC, HCFC, and hydrofluorocarbon (HFC) refrigerants

- Ozone-Depletion Potential (ODP) of CFCs, HCFCs, and HFCs

- Atmospheric effects of the types of refrigerants

- How health and the environment are impacted due to stratospheric ozone depletion

- What is the evidence of stratospheric ozone depletion and the role of CFCs and HCFCs

### (2) Clean Air Act and Montreal Protocol

Here are some important dates and regulations to keep in mind regarding refrigerants and air quality:

- The CFC phaseout date

- The R-22 phaseout date

- Venting during servicing and disposal is prohibited.

- Venting of substitute refrigerants is also prohibited.

- Penalty under the Clean Air Act

- The Montreal Protocol is an international agreement to phase out the production of ozone-depleting substances.

## (3) Section 608 Regulations

- High and low-pressure refrigerants identification

- System-dependent versus self-contained recovery/recycling equipment

- How to identify the equipment covered by the rule (all equipment containing CFCs or HCFCs except motor vehicle air conditioners)

- Reason for third-party certification of recycling and recovery equipment

- The standard for reclaimed refrigerants like Air Conditioning, Heating, and Refrigeration Institute (AHRI) Standard 700-2016

- The restriction on sales

- The prohibition on venting under the Clean Air Act.

## (4) Substitute Refrigerants and oils

- Unavailability of "drop-in" replacements

- Incompatibility of substitute refrigerants with many lubricants used with CFC and HCFC refrigerants

- Incompatibility of CFC and HCFC refrigerants with many new lubricants.

- Identification of lubricants for given refrigerants like with R-134; alkylbenzenes for HCFCs

- The issue of Fractionation (Blends may have a tendency for their components to leak at varying rates.)

## (5) Refrigeration

- Refrigerant states (vapor versus liquid) and pressures at different points of the refrigeration cycle

- How/when cooling occurs

- Refrigeration gauges (color codes, ranges of different types, proper use)

- Leak Detection

## (6) Three R Definitions

- Recover

- Recycle

- Reclaim

## (7) Recovery Techniques

- Need to avoid mixing refrigerants

- Factors affecting the recovery speed (ambient temperature, recycling or recovery equipment size, hose length, diameter, etc.)

### (8) Dehydration Evacuation

- Need to evacuate the system to eliminate air and moisture at the end of service.

### (9) Safety

- What are the risks of exposure to refrigerants?

- What are the personal protective equipment (gloves, goggles, self-contained breathing apparatus (SCBA)-in extreme cases, etc.)

- Reusable (or "recovery") cylinders versus disposable cylinders (ensure former Department of Transportation (DOT) approved, know former's yellow and gray color code, never refill latter)

- Risks of filling cylinders more than 80 percent full

- Use of nitrogen rather than oxygen or compressed air for leak detection.

- Use of pressure regulator and relief valve with nitrogen

### (10) Shipping

- Labels required for refrigerant cylinders (refrigerant identification, DOT classification tag)

Okay, that's the syllabus part for the Core Exam. Now comes important points you should remember from the exam's perspective under each topic of the Core Exam.

# EXAM-SPECIFIC POINTS FOR THE CORE SECTION

## (A) ENVIRONMENTAL IMPACTS

**Ozone Layer:**

- The stratosphere contains a high concentration of ozone (O3) molecules, which serve as a protective layer against the Sun's harmful UV radiation.

- Ozone, made up of three oxygen atoms, is created through chemical reactions from sunlight. If the ozone layer is depleted, it can result in increased UV exposure, which can negatively affect the environment and human health.

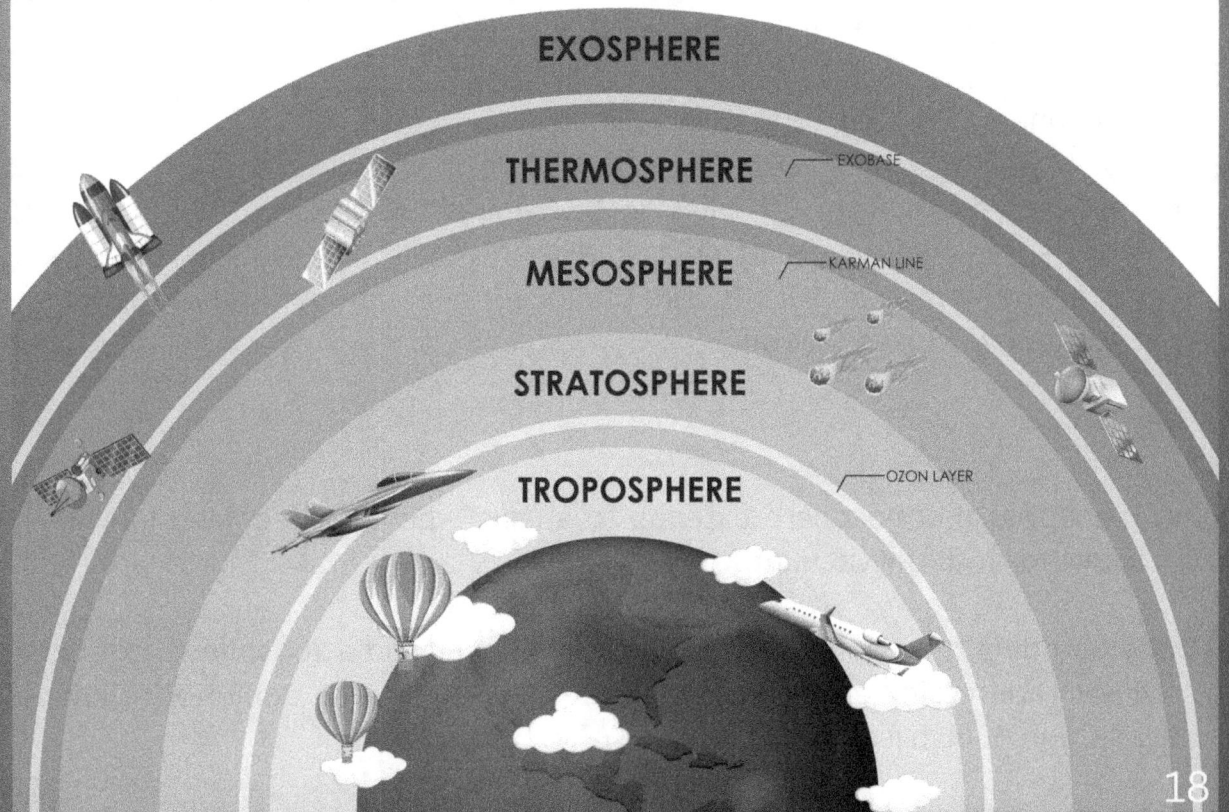

## Atmospheric Layers:

Troposphere:

- The atmosphere stretches from the surface of the Earth up to a distance of 6-7 miles (10-12 km). It contains most of the weather phenomena and atmospheric mass. It is essential for the survival of humans and other living organisms and for regulating climate and weather patterns.

Stratosphere:

- The ozone layer is located 6-30 miles above Earth and causes a temperature increase. It's important because it absorbs UV radiation, which is crucial for protecting life on Earth.

Mesosphere:

- Meteoroids burn like shooting stars when they enter the stratosphere, up to 50 miles (80 km) above the Earth's surface. This usually happens in cold temperatures.

Thermosphere:

- This area spans from 50 miles to 600 miles (80 km to 1,000 km) and experiences high levels of solar radiation absorption, which causes an increase in temperature. It is also the location of various satellites and the International Space Station.

Exosphere:

- The upper thermosphere extends to space, where sparse gas molecules transition into the vacuum.

Understanding the significance of the ozone layer and atmospheric layers is crucial in addressing environmental issues such as ozone depletion, climate change, and human impact.

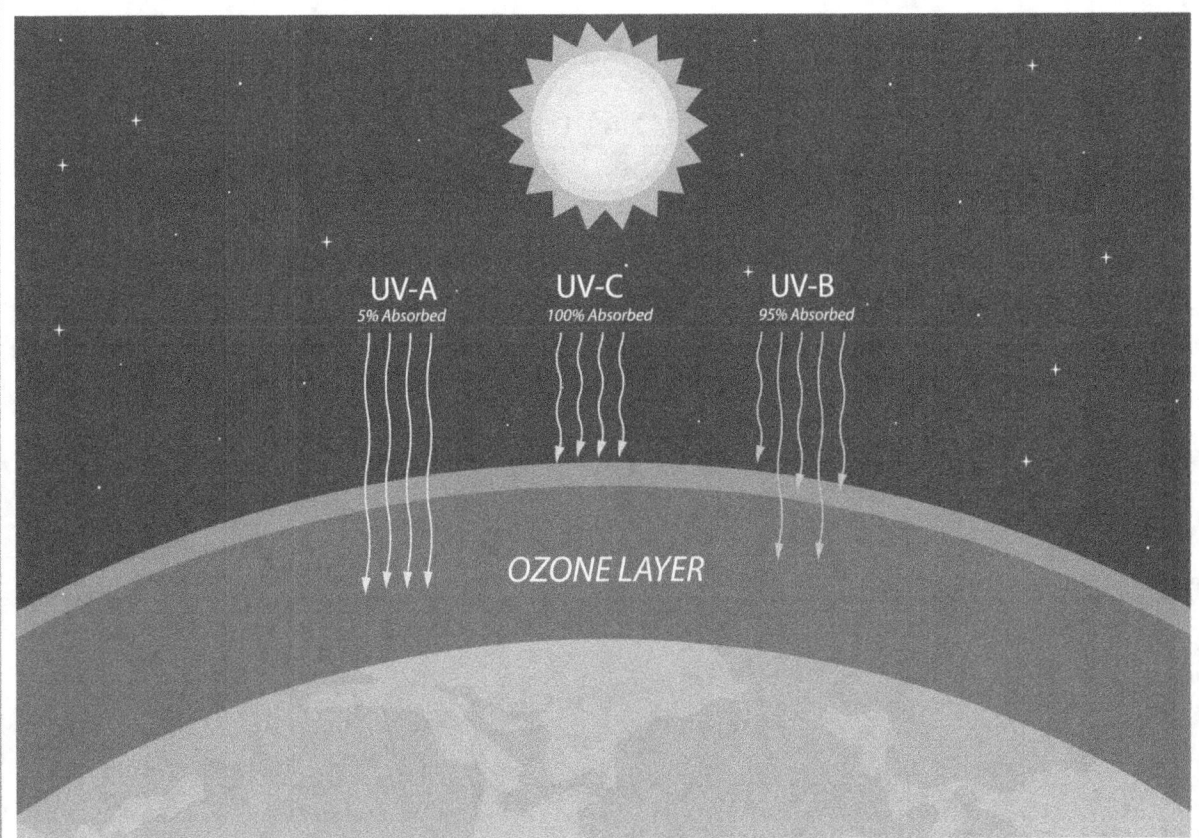

## Importance of the Ozone Layer:

- The ozone layer in the stratosphere protects Earth from harmful UV radiation emitted by the sun.

- The **Ozone Depletion Potential (ODP)** measures how much a substance can harm the ozone layer in the stratosphere.

## Global Impact and Effects:

- The depletion of ozone has a significant impact on the world's population.

- It leads to an increase in skin cancer rates and cataracts.

- It also harms the environment by causing reduced crop yields and damage to marine life.

## How chlorine destroys ozone :

- Ozone in the stratosphere contains three oxygen atoms, while CFC and HCFC molecules contain chlorine. When chlorine reacts with ozone, it creates chlorine monoxide and leaves behind an oxygen molecule. This reaction destroys up to 100,000 ozone molecules per chlorine atom, lasting up to 120 years.

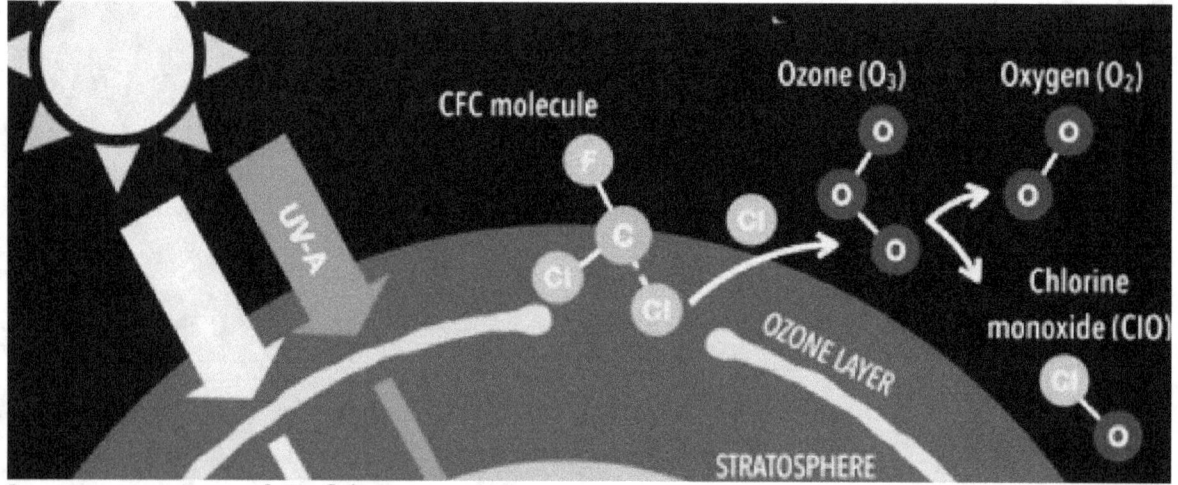

## Controversy and Refrigerant Source:

- There is an ongoing debate about the origin of stratospheric chlorine, but evidence suggests that refrigerants are the main source.

- These refrigerants containing chlorine upset the natural balance and lead to the depletion of the ozone layer.

## Ozone Depletion Mechanism:

- Ozone depletion ($O_3$) in the stratosphere is caused by chlorine atoms in CFCs/HCFCs.

- These chlorine atoms extract oxygen atoms from ozone molecules.

- Additionally, the reaction between chlorine and ozone produces chlorine monoxide (ClO), which attacks more ozone molecules.

**Longevity and Impact of Chlorine:**

- A lone chlorine atom can remain active for as long as 120 years, destroying many ozone molecules.

- The chlorine found in CFCs/HCFCs doesn't dissolve in water or break down, lasting in the atmosphere.

**Different Refrigerant Types:**

- Although HFCs (Hydrogen, Fluorine, Carbon) do not contain ozone-depleting chlorine, they have a high Global Warming Potential (GWP).

- Hydrocarbons (HC) and Hydrofluoroolefins (HFO), on the other hand, have lower GWP when compared to CFCs, HCFCs, and most HFCs.

**EPA Section 608's Objective:**

- The purpose of EPA Section 608 is to prevent the depletion of the ozone layer in the stratosphere.

- The stratosphere, located 6-7 miles to 30 miles above the Earth's surface, contains most ozone concentration.

**Global Warming Potential (GWP) and Refrigerants:**

- GWP compares the warming impacts of gases over time with $CO_2$ as the baseline (GWP=1).

- HFC refrigerants, like R-410A, have significantly higher GWP values than $CO_2$.

- Hydrocarbons (HC) and HFOs have lower GWP values, making them more environmentally friendly.

## Sources of Chlorine in Ozone Depletion:

- The primary sources of chlorine in the atmosphere that contribute to the issue of ozone depletion are CFCs and HCFCs.

- While some individuals may attribute natural events, like volcanic eruptions, to the problem, data indicates that most chlorine in the stratosphere comes from artificial chemicals rather than from natural sources such as volcanoes.

- There is evidence to support this claim, such as the correlation between the rise in fluorine levels - which comes from different sources than chlorine - and the increase in chlorine in the stratosphere. Additionally, the increase in chlorine levels corresponds to the emissions that contribute to ozone depletion.

- Furthermore, research has shown that air samples collected from the stratosphere over erupting volcanoes have less chlorine than CFCs and HCFCs.

- It's unlikely for HCl from volcanoes to reach the stratosphere due to moisture and steam removing chlorine particles. Studies show that only a tiny percentage of stratospheric chlorine comes from natural sources, mostly from substances that damage the ozone layer.

- To prevent further harm to the stratospheric ozone layer, technicians in the United States are required to recover all refrigerants that have an ozone depletion potential (ODP) above zero and a global warming potential (GWP) higher than carbon dioxide (value above 1).

- Although HFCs do not have ODP, their GWP is thousands of times greater than those of hydrofluoroolefin (HFO) and hydrocarbon (HC) refrigerants.

- Using alternative non-ozone-depleting refrigerants and low GWP refrigerants will ultimately eliminate CFCs, HCFCs, and HFCs.

# Identification of CFC, HCFC, and hydrofluorocarbon (HFC) refrigerants

**CFC**
C = CHLORO (Chlorine)
F = FLUORO (Fluorine)
C = Carbon
Common CFC refrigerants include:
R-11, R-12, R-113, R-114, R-115, R-500 and R-502

**HCFC**
H = Hydro (Hydrogen)
C = Chloro (Chlorine)
F = Fluoro (Fluorine)
C = Carbon
Common HCFC refrigerants include R-22, 123 and 124

**HFC**
H = Hydro (Hydrogen)
F = Fluoro (Fluorine)
C = Carbon
Common HFC refrigerants include R-134A

- Different types of refrigerants can be identified by their classification (CFC, HCFC, HFC, HFO, HC) and their ASHRAE number.

- They can also be called "R," followed by their ASHRAE numbers, such as HFC-134a or R-134a.

- ASHRAE has categorized refrigerants into Class A (safest) or B based on their toxicity level for humans.

- Their flammability is denoted by 1 (non-flammable), 2 (low flammability), or 3 (high flammability).

# Ranking Refrigerants for Potential Damage

| Refrigerant Type | Composition | Ozone Depletion Potential (ODP) | Global Warming Potential (GWP) | Flammability | Examples |
|---|---|---|---|---|---|
| CFC | Chlorine, Fluorine, Carbon | Highest potential for ODP | Varies based on type | Non-flammable | R-11, R-12 |
| HCFC | Hydrogen, Chlorine, Fluorine, Carbon | Lower ODP compared to CFCs | Varies (usually low to moderate) | Non-flammable | R-22, R-123 |
| HFC | Hydrogen, Fluorine, Carbon | Low/no potential for ODP | Contributes to global warming | Non-flammable | R-134a, R-410A, R-404A, R-407C, R-422B |
| HFO | Hydrogen, Fluorine, Carbon | No potential for ODP | Minimal effect on global warming | Mildly flammable (A2L) | R-1234yf (for chiller systems) |
| HC | Hydrogen, Carbon | No potential for ODP | Very small effect on global warming | Flammable | Isobutane (R-600a), Propane (R-290) |

# Refrigerant Classification & ASHRAE Safety Group

| Toxicity | Flammability | Classification |
|----------|--------------|----------------|
| High     | High         | B3             |
| Low      | High         | A3             |
| High     | Low          | B2             |
| Low      | Low          | A2             |
| High     | Slight       | B2L            |
| Low      | Slight       | A2L            |
| High     | No           | B1             |
| Low      | No           | A1             |

**Toxicity of Refrigerant:**

Refrigerant toxicity measures how harmful a substance is to people and the environment. High toxicity can cause health problems, while low toxicity is safer.

**Flammability of Refrigerant**:

Refrigerants can be classified as high, low, or slight. High flammability poses a greater safety risk, while low or slight flammability is safer. Handling, storing, and using refrigerants safely is essential to avoid fire hazards.

To minimize harm to people and the environment, refrigerants with low toxicity and flammability are preferred for certain applications.

# Ranking Refrigerants for Potential Damage

- Refrigerants can harm the atmosphere, causing ozone depletion and global warming.

- CFC and HCFC refrigerants are bad for ozone depletion, but HFC refrigerants are not.

- Ozone in the Stratosphere is good, as it acts as a UV filter, while ozone in the Troposphere is terrible as a pollutant.

- Chemicals in HVACR have depleted the ozone layer, creating a hole. Ozone in the Troposphere is a pollutant and contributes to global warming.

- Stratospheric ozone filters harmful UV sunlight. Too much UV exposure can weaken the immune system and increase cancer risks.

- The greenhouse gas effect causes global warming by trapping heat near the earth.

- Evidence shows that artificial Chlorine from CFCs and HCFCs is the cause of ozone depletion.

- Fluorine, which does not naturally occur with Chlorine, is present in the Stratosphere, indicating that using artificial refrigerants contributes to the issue.

**IMPACT OF OZONE DEPLETION**

- Increased incidence of certain skin cancers and cataracts
- Suppression of the immune system
- Decreased crop yields and damage to marine organisms
- Increased formation of ground-level Ozone
- Increased weathering of outdoor plastics
- Increased temperature of the earth
- Advanced cases of skin cancer
- Increased numbers of cataracts in the eyes
- Increased ground-level ozone
- Crop and vegetation loss
- Reduced marine life

# (B) CLEAN AIR ACT AND MONTREAL PROTOCOL

- In 1963, the **Clean Air Act** was introduced to provide funds for researching and removing air pollution. However, it wasn't until 1970, when a more robust version of the Act was passed, that the federal government took comprehensive action to combat air pollution across the entire United States.

- That same year, the **Environmental Protection Agency (EPA)** was established by Congress, with the primary responsibility of enforcing the law. The EPA has since implemented various Clean Air Act programs to reduce air pollution nationwide.

- In 1990, the Clean Air Act underwent a significant revision and expansion by Congress, granting the EPA more extensive power to enforce regulations to reduce air pollutant emissions.

- The amendments also prioritized cost-effective methods to decrease air pollution. The program included a plan for gradually discontinuing the production and use of substances that destroy the ozone layer and capturing and ultimately eliminating CFCs.

- However, the program did not authorize the EPA to compel the removal of high GWP refrigerants. The mandate solely focused on eliminating refrigerants that deplete the ozone layer.

- During the Fourth Meeting of the **Montreal Protocol in 1992**, participating parties adopted several measures to address the use of hazardous chemicals. These measures included expediting the elimination of CFCs, halons, carbon tetrachloride, and methyl chloroform and adding HCFCs and methyl bromide to the list of regulated chemicals. By 1994, halons had been entirely phased out by these measures.

- In 1995, the manufacturing and distribution of new CFCs, halogenated CFCs, carbon tetrachloride, and methyl chloroform ceased. Later, in 2010, the production and sale of new HCFCs also ended.

- Apart from phasing out harmful refrigerants, the Clean Air Act also made releasing refrigerants into the air illegal and empowered the EPA to set guidelines for safe refrigerant recovery.

**CFC Phase-out Date**

The CFC phaseout protects the ozone layer from harmful UV radiation by reducing the usage of substances that cause depletion. EPA has a plan to decrease usage and preserve the ozone layer gradually.

The CFC phaseout dates for the US are as follows:

- Production and import: January 1, 1996

- Use in new equipment: January 1, 2010

- Service and repair: January 1, 2020

**R-22 Phase-out Date**

- R-22 is a type of refrigerant commonly used in air conditioners and refrigerators but it also harms the ozone layer.

- Because of this, it is being phased out under the Montreal Protocol. The phaseout of R-22 is happening in stages to give time for developing and adopting alternative refrigerants.

- After the phaseout, technicians must use alternative refrigerants when servicing and repairing equipment that previously used R-22. Alternative refrigerants are available, but some may be more expensive than R-22.

The R-22 phaseout dates for the US:

- Production and import: January 1, 2020

- Use in new equipment: January 1, 2024

- Service and repair: January 1, 2030

# Maximum Penalty under Clean Air Act

Sources:

https://www.lexology.com/library/detail.aspx?g=0ced5e72-9528-4de9-bbd0-85f870b3ba1

https://complynet.com/resources/articles/epa-increases-penalties-7.745-2023-0

https://orr-reno.com/the-cost-of-noncompliance-osha-dot-epa-penalties-increase-for-2023/

| Type of Violation | Maximum Penalty (Individuals) | Maximum Penalty (Organizations) | Maximum Penalty (Government Agencies) |
|---|---|---|---|
| Knowingly and willfully violating | $446,456 | $1,174,680 | $10,000 |
| Negligent violations | $55,808 per day per violation | $117,468 per day per violation | $10,000 per day per violation |

|  | 2022 Penalties | 2023 Penalties |
|---|---|---|
| Clean Air Act Daily: Maximum (per violation): | $51,796 – $109,024<br>$414,364 | $55,808 – $117,468<br>$446,456 |
| Clean Water Act Daily: | $23,989 – $59,973 | $25,847 – $64,618 |

Under the 2023 rule, the new maximum EPA civil penalties are:

**Clean Air Act** (Daily)          $55,808 – $117,468

Maximum (per violation)          $446,456

# Venting

- **Venting prohibition:** The Clean Air Act (CAA) has a regulation called the Venting Prohibition, which forbids intentionally releasing refrigerants, such as CFCs, HCFCs, and HFCs.

- **De minimis releases:** Small, unintended refrigerant emissions during recovery or regular appliance operation are known as de minimis releases. These releases are allowed under the venting prohibition.

- **Nitrogen:** It is possible to release nitrogen without breaking the venting prohibition as long as it is not utilized to pressurize a system to detect leaks.

- **Appliance disposal:** Before disposing of an appliance that contains refrigerants, it's important to recover the refrigerants first. The individual responsible for disposing of the appliance must remove the refrigerants before final disposal.

**Release allowances and exemptions**: A few exceptions exist to the venting prohibition. These exceptions include:

Here are some instances where emissions of refrigerants are allowed:

- "De minimis" releases (Small, unintended refrigerant emissions)

- Emissions during normal system operation

- Use of a trace gas mixture for leak detection

- Repair of significant leaks for systems with more than 50 pounds of charge

- Specific releases of CFCs or HCFCs used for holding charges or purging hoses

- Exempt refrigerants like carbon dioxide or R-290 propane.

**Substitute refrigerants:** It has been illegal to release CFC and HCFC refrigerants since November 1995 intentionally. This also applies to HFCs, as they have a high potential for global warming rather than ozone depletion.

Here are some additional points to keep in mind:

- It is important to note that the prohibition on venting refrigerants applies to all types of equipment, such as air conditioners, refrigerators, and chillers.

- This prohibition covers all servicing activities, including installation, maintenance, and repair.

- While there are some exemptions, they are limited in scope. Knowing the venting prohibition regulations is crucial as a technician working with refrigerant equipment.

- For further information on these regulations, please visit the EPA website.

## (C) SECTION 608 REGULATIONS

- The EPA regulates the recycling of ozone-depleting refrigerants under the Clean Air Act. Venting refrigerants from any equipment is illegal, including recycling and storage cylinders.

- The EPA regulations significantly reduce the emission of ozone-depleting refrigerants. They require service technicians to reduce refrigerant emissions by following certain practices, establishing certification programs for equipment and off-site reclaimers, and establishing a technician certification program.

- Certified techs must buy refrigerants. No certification is needed for tasks that don't release refrigerant, like changing a capacitor or rewiring a circuit.

- Section 608 mandates individuals servicing or disposing of air conditioning and refrigeration equipment to comply with recovery requirements and acquire refrigerant recovery and recycling equipment.

- MVAC service technicians are governed by Section 609 of the Clean Air Act and must obtain a Section 609 MVAC certification.

- Maintaining adequate refrigerant supplies for service calls after production bans, preventing the venting of refrigerants to the atmosphere, and preventing stratospheric ozone depletion necessitate standards for recovering refrigerants.

**Definition/identification of high and low-pressure refrigerants**

| Pressure Classification | Boiling Point at Atmospheric Pressure | Pressure at 104°F condensing temperature | Examples |
|---|---|---|---|
| Low-pressure | Above 10°C (50°F) | 30 psig or lower | CFC-11, HCFC-123, HFC-245fa, HFO-1233zd |
| Medium-pressure | -50°C to 10°C (-58°F to 50°F) | 30 psig to 155 psig | CFC-12, HCFC-124, HCFC-409A, HFC-134a, HC-600a, HFO-1234yf, HFO-1234ze |
| High-pressure | -50°C to 10°C (-58°F to 50°F) | 155 psig to 340 psig | HCFC-22, HFC-404A, HFC-407C, HFC-422B, HFC-422D, HFC-410A, HC-441A, R-717 |
| Very high-pressure | Below -50°C (-58°F) | Over 340 psig | $CO_2$ (R-744) |

- According to the regulations for EPA 608, refrigerants are classified as high-pressure, very high-pressure, or low-pressure based on their pressure at a temperature of 104°F while condensing.

This classification plays a crucial role in identifying the appropriate recovery and recycling equipment to service refrigerant-containing equipment.

**Definition of system-dependent versus self-contained recovery/recycling equipment**

| Feature | System-dependent recovery | Self-contained recovery |
|---|---|---|
| Requires | Compressor of the refrigeration system | Its own compressor |
| Typically used for | Small appliances | Wider range of appliances, including chillers and air conditioners |
| Cost | Less expensive | More expensive |
| Efficiency | Not as efficient | More efficient |
| Advantages | Portable, less expensive, less complex | Can be used for a wider range of appliances, more efficient |
| Disadvantages | Can only be used for small appliances, less efficient | More expensive, bulky, not as portable |

**System-dependent recovery equipment**

- It uses the refrigeration system's compressor to remove the refrigerant from the system.

- This type of equipment is typically used for small appliances, such as refrigerators and freezers.

**Self-contained recovery equipment**

- It is equipment that has a compressor and pump.

- This type of equipment can be used for a broader range of appliances, including larger systems such as chillers and air conditioners.

**Identification of equipment covered by the rule**
*(all air-conditioning and refrigeration equipment containing CFCs or HCFCs except motor vehicle air conditioners)*

Equipment Covered by the Section 608 Regulations:

The regulations of EPA Section 608 apply to every air-conditioning and refrigeration equipment that contains CFCs or HCFCs, regardless of whether it is used in homes, businesses, or industrial settings. It is important to note that the Section 608 regulations do not cover motor vehicle air conditioners.

The specific equipment covered by the Section 608 regulations is listed :

- Air conditioners

- Heat pumps

- Refrigeration systems

- Refrigerant appliances

- Fire suppression systems

- Foam blowing machines

- Solvent recovery units

- Metered dispensing systems

- The Section 608 regulations also apply to any equipment modified to contain CFCs or HCFCs.

## Need for third-party certification of recycling and recovery equipment

The EPA requires a third-party certification for equipment used to recycle CFCs or HCFCs to ensure safe and responsible handling.

The certification process includes testing the equipment to meet the EPA's performance standards. The equipment must also be labeled with the certification number.

The EPA has approved several laboratories and organizations to certify recycling and recovery equipment. These organizations are listed on the EPA's website.

### Importance of Third-Party Certification

Third-party certification of recycling and recovery equipment is essential for several reasons:

- This process ensures that the equipment can safely recover and recycle refrigerants in an environmentally responsible manner.

- It protects technicians from exposure to hazardous refrigerants and prevents the release of refrigerants into the atmosphere.

- Additionally, it contributes to the conservation of natural resources.

By requiring third-party certification of recycling and recovery equipment, the EPA is helping to protect the environment and public health.

# The standard for reclaimed refrigerant
*(Air Conditioning, Heating, and Refrigeration Institute (AHRI) Standard 700-2016)*

## The Standard for Reclaimed Refrigerant (ARI 700-2016)

The AHRI Standard 700-2016 establishes purity specifications, verifies composition, and specifies testing methods to ensure the acceptability of fluorocarbon refrigerants (whether new, reclaimed, or repackaged) for use in refrigeration and air-conditioning products within the scope of AHRI, both new and existing.

The purpose of the standard is to guarantee that the recovered refrigerant is of sufficiently high quality to be utilized in refrigeration and air-conditioning systems.

The standard outlines the acceptable levels of various contaminants in the recovered refrigerant, including:

- Moisture

- Non-condensable gases

- Contaminants that can cause corrosion or degradation of system components

The standard mandates the utilization of specific testing techniques to examine reclaimed refrigerant for the presence of contaminants.

## Importance of the Standard

AHRI Standard 700-2016 is vital for preserving the quality of recycled refrigerants. It prevents the release of contaminated refrigerants and ensures technician safety.

The standard guarantees reliable refrigeration and air-conditioning systems, and superior quality recycled refrigerants can be safely used by following it.

## The sales restriction

- To protect the ozone layer and prevent climate change, only certified technicians can purchase chlorofluorocarbons (CFCs) and hydrochlorofluorocarbons (HCFCs) as per the EPA Section 608 regulations.

- This rule has been in place since November 14, 1994, and applies to all containers of CFCs and HCFCs, regardless of size. However, there are exceptions to this rule.

- For instance, technicians certified under Section 609 of the Clean Air Act, which governs motor vehicle air conditioning, can purchase refrigerants in containers of any size.

- The sales restriction also does not apply to refrigerants in fully assembled and sealed appliances.

- This means you do not need to be EPA-certified to purchase a refrigerator, even though it has refrigerant.

The sales restriction is essential for several reasons:

- It helps to ensure that only qualified individuals handle CFCs and HCFCs.

- It helps to prevent the release of these refrigerants into the atmosphere.

- It helps to protect the ozone layer and the environment.

By following the sales restriction, technicians can help to protect the environment and public health.

- According to EPA Section 608 regulations, EPA certification is mandatory for all technicians who handle CFCs and HCFCs.

- Various resources on the EPA's website can assist technicians in understanding the Section 608 regulations and the proper handling of refrigerants. These resources are readily available for use.

## The Clean Air Act prohibition on venting

It is against the Clean Air Act to intentionally release refrigerants into the atmosphere. This applies to refrigerants released from appliances, equipment, and recovery cylinders.

There are a few exceptions to the venting prohibition, such as:

- Leaks that are unavoidable during the normal course of servicing equipment

- The release of small amounts of refrigerant during the evacuation of an appliance

- The release of refrigerant during the disposal of appliances and equipment

- The EPA can fine technicians who violate the venting prohibition.

## (D) SUBSTITUTE REFRIGERANTS AND OILS

### Absence of "drop-in" replacements

- It is important to note that no direct CFC and HCFC refrigerants replacements exist. Therefore, it is necessary to replace both the refrigerant and the oil in the system when making a replacement.

- Using a new refrigerant with old oil can lead to issues like compressor damage due to potential incompatibility.

- If you want to replace R-22 with R-134a in a system that initially used CFC and HCFC systems with mineral oil, you must be aware of their incompatibility. To ensure that the system runs smoothly, you must flush the system and replace the mineral oil with a synthetic oil compatible with R-134a.

## Incompatibility of Substitute Refrigerants with Lubricants

- It's important to note that certain substitute refrigerants, like HFC-134a, may not work well with the mineral oils typically used in CFC and HCFC systems.

- If you're replacing a CFC or HCFC refrigerant with HFC-134a, it's necessary to flush the system and replace the oil.

- On the other hand, newer lubricants may not be compatible with CFC and HCFC refrigerants. For example, esters are an excellent choice for lubricating R-134a, while alkylbenzenes work better with HCFCs.

## Fractionation Problem in Refrigerant Blends

- Blended refrigerants, like R-410A, are susceptible to a phenomenon known as fractionation.

- This occurs when the various elements of the mixture escape from the system at varying speeds.

- If left unchecked, this could result in a decline in performance and potentially harm the system.

| Refrigerant | Compatible Oil |
| --- | --- |
| R-12 | Mineral oil |
| R-22 | Mineral oil |
| R-134a | Synthetic oil (POE or PAG) |
| R-410A | Synthetic oil (POE or PAG) |
| R-404A | Synthetic oil (POE or PAG) |

# Refrigerant Oils:

## Refrigerant Oils and Type

The usage of refrigerants and oils varies depending on their type, whether new, old, or blended.

**Mineral Oils**: These include paraffin-based, naphthene-based, and mixed oils.

**Synthetic Oils**: Types like silicate ester, silicone, neo-pentyl ester, dibasic acid ester, PAG, AB, and POE.

- Synthetic oils should be stored in metal containers.
- Ester oils are often used with alternative refrigerants and are compatible with mineral oils and existing system components.

## Oil Types and Use

| Oil Type | Abbreviation | Use |
| --- | --- | --- |
| Mineral Oil | MO | CFC refrigerant systems, some HFC refrigerant systems |
| Alkylbenzene | AB | R-22 and other refrigerant systems |
| Polyolester | POE | HFC refrigerant systems, some blends |
| Polyalkylene Glycol | PAG | R-134a automotive systems, some blends |
| Polyalphaolefin | PAO | R-717 (ammonia) refrigeration systems, some blends |

## Properties and Characteristics of Refrigerant Oils

Refrigerant oil must possess the following properties for effective performance:

**Miscibility with refrigerants**: The oil and refrigerant must blend without separating for optimal performance.

This allows the oil to lubricate all the moving parts of the compressor and other components in the system efficiently.

**Lubrication properties**: The oil must be able to lubricate the compressor's moving parts and other system components, even in refrigerant dilution, to avoid any damage or deterioration to the system components.

**Electrical insulating properties:** The oil must not conduct electricity, which could cause a shock hazard.

**Oxidation stability:** The oil must have oxidation resistance to prevent it from breaking down and forming sludge.

This is necessary to ensure the oil does not clog the system and lead to other issues.

**Low-temperature properties**: The oil must remain fluid at low temperatures to lubricate the compressor and other components.

Low-vapor pressure: The oil must have a low vapor pressure so that it does not evaporate quickly and cause the system to lose refrigerant.

**Chemical compatibility:** The oil must be compatible with the refrigerant and other components in the system.

**Cost**: The oil must be cost-effective.

## Temperature Glide and Blend Behavior

Certain refrigerants are combined with other refrigerants to attain particular characteristics, resulting in what is known as zeotropic blends.

One type of zeotropic blend, EMPZeotropic blend, includes temperature glide, which is the variance in temperature between the blend's dew point and bubble point.

The dew point refers to the temperature at which the refrigerant vapor transforms into liquid. In contrast, the bubble point refers to the temperature at which the refrigerant liquid transforms into vapor.

Temperature glide can lead to issues in refrigeration systems, such as uneven refrigerant distribution and reduced efficiency.

## Zeotropic and Azeotropic Refrigerants

There are two types of refrigerants: zeotropic and azeotropic.

Zeotropic refrigerants are blends that maintain constant pressure while changing composition and saturation temperatures. This causes them to exhibit temperature glide.

On the other hand, azeotropic refrigerants are blends that contain fluids with the same boiling temperatures and behave like a single refrigerant. As a result, they do not exhibit temperature glide.

## Charging Zeotropic and Azeotropic Refrigerants

Charging zeotropic refrigerants as a liquid is important to ensure even distribution throughout the system.

On the other hand, azeotropic refrigerants can be charged either as vapor or liquid.

# (E) REFRIGERATION

**Basics of Refrigeration Principle :**

- Cooling (refrigeration or air conditioning) involves removing heat from a lower-temperature area and releasing it into a higher-temperature location.

- Heat naturally moves from hot to cold, similar to water flowing downhill.

- Achieving heat flow from cold to hot requires a heat pump, similar to using a water pump to move water uphill.

- In the HVAC/R industry, all refrigeration and air conditioning systems are essentially heat pumps.

- Specifically, a heat pump in HVAC/R is a vapor-compression cooling system with a reversible valve.

- This valve allows the system to provide interior cooling during hot ambient temperatures and internal heating during cold ambient temperatures.

**Components of Refrigeration system**

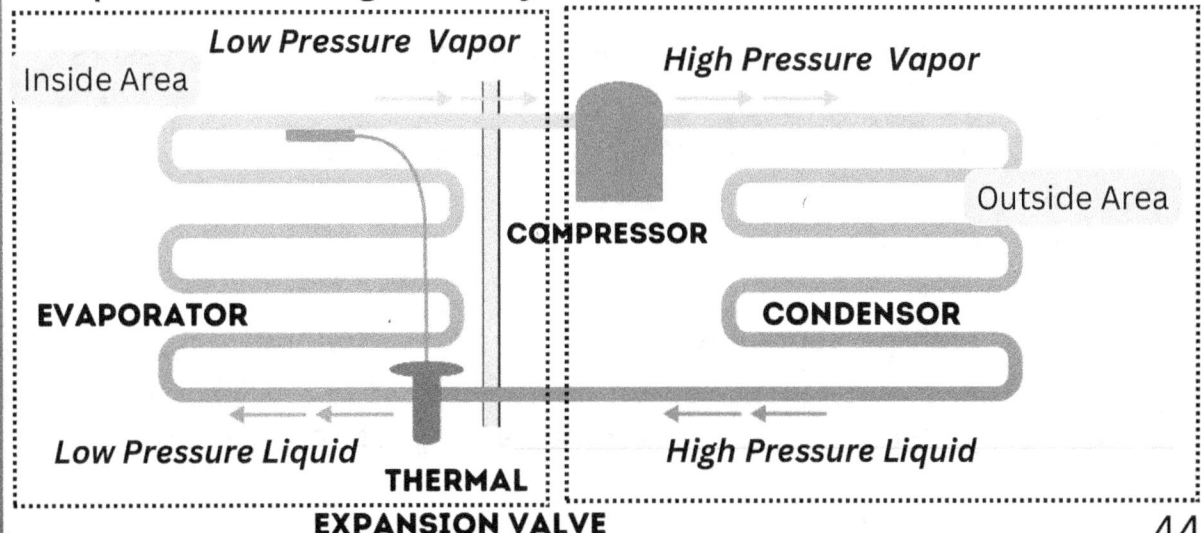

| Component | State | Pressure | In | Out |
|---|---|---|---|---|
| Compressor | Vapor | Low | Low-pressure vapor | High-pressure vapor |
| Condenser | Vapor | High | High-pressure vapor | High-pressure liquid |
| Expansion Valve | Liquid | High | High-pressure liquid | Low-pressure liquid |
| Evaporator | Liquid | Low | Low-pressure liquid | Low-pressure vapor |

- **Compressor:**

Converts **low-pressure vapor to high-pressure vapor.**

Types of compressors:

(a) Small Compressors: Reciprocating, rotary, and scroll compressors

(b) Large Compressors: centrifugal and screw compressors.

- **Condenser:**

(a) Converts **high-pressure vapor to a high-pressure liquid.**

(b) Heat is rejected from the refrigerant, causing it to condense.

- **Expansion Valve or Throttling Device:**

(a) Reduces pressure to lower saturation temperature.

(b) Allows refrigerant to evaporate or boil in the evaporator.

(c) Draws heat into the refrigerant during evaporation.

- **Evaporator:**

(a) Converts low-pressure two-phase mixture (liquid and vapor) into the all-vapor refrigerant.

(b) Heat is drawn into the refrigerant during evaporation, providing cooling.

**Summary:**

The common refrigeration systems consist of four major components that work together:

- The compressor raises vapor pressure.

- The condenser converts high-pressure vapor to liquid by releasing heat.

- The expansion valve lowers pressure, enabling evaporation in the evaporator.

- The evaporator transforms the liquid-vapor mixture to an all-vapor state, drawing heat and cooling.

# Refrigeration Cycle Process

The refrigeration cycle process involves transferring heat from one location to another. This mechanism is utilized in various cooling devices such as refrigerators, freezers, and air conditioners.

The refrigeration cycle has four main steps:

- **Compression:** The compressor takes in a low-pressure vapor refrigerant and compresses it into a high-pressure, hot vapor.

- **Condensation**: The condenser cools the high-pressure, hot vapor and condenses it into liquid.

- **Expansion:** The expansion valve reduces the pressure of the liquid refrigerant, causing it to vaporize.

- **Evaporation**: The evaporator absorbs heat from the surrounding area and evaporates the refrigerant.

Here are the details of each step:

**Compression:** The compressor is powered by an electric motor to take in low-pressure vapor refrigerant and compress it into high-pressure and hot vapor.

This process results in the refrigerant heating up due to the work done by the compressor.

**Condensation:** The condenser usually has copper or aluminum tubes and fins.

The refrigerant in a high-pressure and hot vapor state flows through the condenser's coils.

The fins assist in dispersing the heat from the refrigerant into the nearby air. Ultimately, the refrigerant condenses and converts back into a liquid form.

**Expansion**: The expansion valve is a device that reduces the pressure of the liquid refrigerant. This causes the refrigerant to vaporize. The expansion valve is typically a small orifice or capillary tube.

**Evaporation:** The evaporator usually comprises copper or aluminum tubing with fins. The low-pressure refrigerant in vapor form flows through the evaporator coils, while the fins aid in absorbing heat from the nearby environment. The refrigerant undergoes vaporization, taking in heat from its surroundings and bringing about a cooling effect.

The refrigeration cycle is a self-contained system where the refrigerant is recycled continuously. Only when there is a leak in the system is refrigerant added.

Refrigerant migration is when refrigerant moves to the coldest part of the system during shut down or start-up of the compressor. To prevent this, a crankcase heater is typically utilized.

This heater is placed on the outside of the compressor and is activated whenever the compressor isn't in use. It warms up the compressor crankcase to prevent refrigerant from migrating.

**How/When Cooling Occurs**

The process of cooling takes place in the evaporator as the refrigerant absorbs heat from the area being cooled and transforms it into vapor. The vapor then carries the absorbed heat to the condenser, which is released to the outside air.

The amount of cooling that occurs depends on the following factors:

- The temperature of the space being cooled

- The temperature of the outside air

- The amount of refrigerant in the system

- The efficiency of the refrigeration system

# Refrigeration Gauges
## (color codes, ranges of different types, proper use)

A refrigeration gauge is a device used to measure the pressure of refrigerant in a refrigeration system.

There are two primary refrigeration gauges: manifold and single gauges.

- **Manifold gauges:** A manifold gauge is a pair of gauges linked to a manifold. One gauge tracks the low-pressure side of the system, while the other measures the high-pressure side.

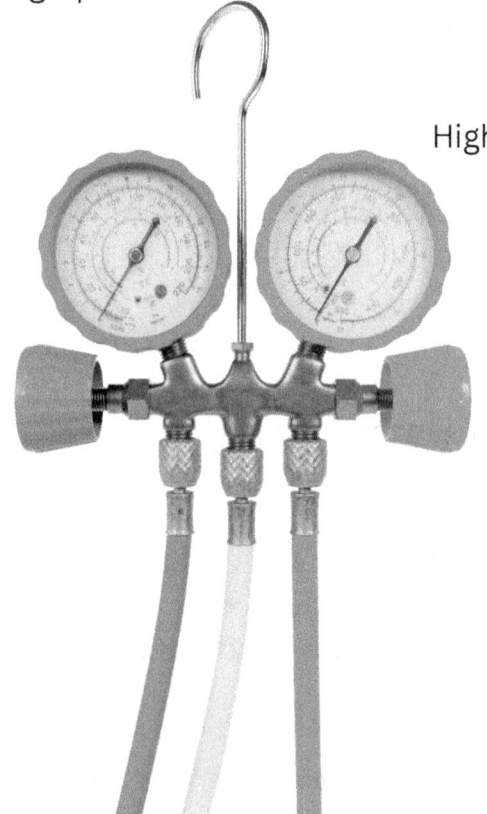

**2 Valve 3 Hose Manifold Gauge**

- **Single gauges**: Single gauges measure the pressure on either the low-pressure or high-pressure side of the system.

**Refrigeration gauges are used to:**

- Diagnose problems with refrigeration systems
- Charge refrigeration systems with refrigerant
- Evacuate refrigeration systems
- Recover refrigerant from refrigeration systems

**Color Codes:**

Refrigeration gauges use color coding to identify the refrigerant they are intended to measure. Here are the most prevalent color codes:

- **Blue**: R-12, R-22, R-134a, and other common refrigerants
- **Red**: R-410A and other high-pressure refrigerants
- **Yellow**: R-502 and other low-pressure refrigerants
- **Black**: Used for vacuum readings
- **Green**: Used for manifold pressure readings

**Ranges of Different Types:**

Different ranges of refrigeration gauges are available, depending on the specific refrigerant that needs to be measured. Here are the most commonly used ranges:

- **Low-pressure gauges**: Range from 0 to 300 psi
- **High-pressure gauges**: Range from 0 to 1000 psi
- **Vacuum gauges**: Range from 30 to 3000 microns

**Proper Use:**
To use refrigeration gauges safely, always follow the manufacturer's instructions. Here are some general safety tips to keep in mind when using them:

- Always wear safety glasses when using refrigeration gauges.

- Never use a gauge that is damaged or leaking.

- Before reading, please make sure the gauge is securely connected to the refrigerant line.

- Refrain from over-pressurizing the gauge.

- Release the pressure from the gauge before disconnecting it from the refrigerant line.

## Table for color codes for refrigeration gauges

| Color | Refrigerant | Range | Manifold or Single Gauge | Notes |
|---|---|---|---|---|
| Blue | R-12, R-22, R-134a, and other common refrigerants | 0 to 300 psi | Manifold gauge | Most common color code for refrigeration gauges. |
| Red | R-410A and other high-pressure refrigerants | 0 to 1000 psi | Manifold gauge | Used for high-pressure refrigerants. |
| Yellow | R-502 and other low-pressure refrigerants | 0 to 300 psi | Manifold gauge | Used for low-pressure refrigerants. |
| Black | Used for vacuum readings | 30 to 3000 microns | Single gauge | Used to measure vacuum levels in refrigeration systems. |
| Green | Used for manifold pressure readings | 0 to 300 psi | Manifold gauge | Used to measure the pressure on both the low-pressure and high-pressure sides of a refrigeration system. |

**How to Use a Gauge Manifold Set**

Connect the gauges to the refrigeration system via the provided hoses using a gauge manifold set.

Make sure to connect the high-pressure gauge to the high-pressure port on the system and the low-pressure gauge to the low-pressure port.

The third port may be left open or linked to any of the following:

- a recovery device
- an evacuation pump or a charging device.

After connecting the gauges, you can obtain readings by simply checking them.

The high-pressure gauge displays the pressure on the high-pressure end of the system, while the low-pressure gauge displays the pressure on the low-pressure end of the system.

Here is a more detailed explanation of the gauge manifold set:

- **Compound Gauge**: Compound gauges measure both high and low-pressure refrigerant sides in psig (pounds per square inch gauge). They have a blue gauge for low pressure and a red gauge for high pressure. They also measure vacuum in inches of mercury.

- **High-pressure Gauge**: The red gauge measures high pressure on the side closer to the compressor in a refrigeration system.

- **Center Port**: The center port is a third port on the manifold that can be used for various purposes. It can be used to connect a recovery device, an evacuation pump, or a charging device. The center port is typically colored yellow.

- **Low-Loss Fittings**: EPA regulations require that the hoses connected to the gauge manifold set be equipped with low-loss fittings. Low-loss fittings minimize the amount of refrigerant lost when the hoses are disconnected.

## Leak Detection

Detecting and fixing leaks in refrigeration systems is crucial to prevent poor performance, early component failure, and environmental damage. Common methods for leak detection include various techniques.

- **Pressure testing** involves pressurizing the system with an inert gas, such as nitrogen, and then looking for leaks with a soap solution or electronic leak detector.

- **Visual inspection** involves looking for signs of leaks, such as oil stains, frost, or refrigerant gas escaping from the system.

- **Halide torch testing** involves using a halide torch to create a flame that will change color when it comes into contact with refrigerant gas.

- **Dye injection** involves injecting a fluorescent dye into the system and then using a UV light to look for leaks.

Fix refrigerant leaks promptly. EPA requires fixing leaks in systems containing 50+ pounds within 30 days of detection.

Here are some **tips for preventing leaks** in refrigeration systems:

- Use approved equipment and procedures.

- Inspect the system regularly for signs of leaks.

- Tighten all connections regularly.

- Use a leak detector to check the system for leaks before and after servicing.

- Repair leaks immediately.

By following these tips, you can help to prevent leaks in your refrigeration systems and keep them running smoothly and efficiently.

I want to share some essential guidelines from the EPA regarding leak detection and repair in refrigeration systems.

- If a system contains 50 pounds or more of refrigerant and a leak is detected, it must be repaired within 30 days.

- While leaks in systems containing less than 50 pounds of refrigerant don't require repair, it's recommended to do so.

- The EPA mandates using approved equipment and procedures for leak detection and repair.

- Only qualified technicians are permitted to perform leak repairs and must maintain records of all repairs.

| Leak Detection Method | Description | Advantages | Disadvantages |
|---|---|---|---|
| Pressure Testing | The system is pressurized with an inert gas, such as nitrogen, and then checked for leaks with a soap solution or electronic leak detector. | This method is effective for detecting leaks of all sizes. | It can be destructive to the system if the pressure is too high. |
| Visual Inspection | The system is visually inspected for signs of leaks, such as oil stains, frost, or refrigerant gas escaping from the system. | This method is easy to perform and does not require any special equipment. | It can be difficult to identify small leaks. |
| Halide Torch Testing | A halide torch is used to create a flame that will change color when it comes into contact with refrigerant gas. | This method is effective for detecting leaks of all sizes. | It can be dangerous if not used properly. |
| Dye Injection | A fluorescent dye is injected into the system and then checked for leaks with a UV light. | This method is effective for detecting leaks of all sizes. | It can be messy and time-consuming. |

## (F) THREE R DEFINITIONS

| R | Definition | Requirements |
|---|---|---|
| **Recover** | The process of removing refrigerant from a system in any condition and storing it in an external container without testing or processing it in any way. | Required for all systems that contain regulated refrigerants. |
| **Recycle** | The process of cleaning refrigerant for immediate reuse by separating the oil from the refrigerant and removing moisture and acidity from the refrigerant by use of products like filter driers. | Optional process that can be performed on recovered refrigerant. |
| **Reclaim** | The process of processing refrigerant to the level of new product specifications as determined by chemical analysis. | Most thorough process that results in refrigerant that is indistinguishable from new refrigerant. Reclaimed refrigerant must meet the standard set forth in ARI 700 before it can be resold. |

Technicians must follow the Three R's of refrigeration - **Recover, Recycle, and Reclaim** - to ensure refrigerants' safe and eco-friendly handling. These steps are crucial for maintaining a healthy environment.

**Recover**

To begin, it is necessary to extract the refrigerant from the system. This process involves using a recovery device, which is a machine designed to remove the refrigerant from the system and store it in an external container. This must be done for all systems that contain regulated refrigerants like HFCs and HCFCs.

**Recycle**

Recycled refrigerant is safe for use in other systems. To achieve this, impurities are eliminated through techniques such as filtering.

**Reclaim**

Refrigerant reclamation purifies and eliminates contaminants to meet new product specifications. Reclaimed refrigerant can be sold as new.

**Recovery Devices**

- There are two types of recovery devices: system-dependent and self-contained.

- System-dependent devices use the appliance's components to capture refrigerant and are used for smaller appliances.

- Self-contained devices have their own mechanisms and are used for larger appliances and systems with high refrigerant charges.

Technicians working with refrigerants must be EPA-certified and use approved equipment following established recovery, recycling, and reclamation guidelines.

# (G) RECOVERY TECHNIQUES

**Need to avoid mixing refrigerants.**

It is important never to mix refrigerants as it can lead to dangerous and unpredictable outcomes. This can include decreased efficiency, increased wear and tear on the system, and potentially forming explosive compounds.

**Factors affecting the recovery speed**

Several factors can influence the speed of recovery, such as:

- **Ambient temperature**: When the temperature drops, the recovery process slows down. This happens because the refrigerant becomes less volatile in colder weather, which makes it harder to vaporize and eliminate from the system.

- **Size of the system and recovery equipment**: Larger systems and recovery equipment take longer to complete recovery due to more refrigerant needing removal and equipment needing to handle an increased flow rate.

- **Length and diameter of the hoses:** If the hoses used for refrigerant recovery are longer and narrower, the recovery process will take longer. This is because the pressure drop will increase, leading to additional work for the recovery equipment to transfer the refrigerant through the hoses.

**Other factors that can affect the speed of recovery include:**

- **The type of refrigerant being recovered:** Various refrigerants have unique boiling points and densities. These properties can impact the speed of their recovery.

- **The condition of the refrigerant system:** A system leaking refrigerant will take longer to recover than one not leaking.

- **The presence of leaks in the system:** Leaks in the system can cause refrigerant to escape into the atmosphere, reducing the amount of refrigerant that can be recovered.

## Identifying the Refrigerant

It's important to identify the type of refrigerant in the system before starting a refrigerant recovery procedure.

Each type of refrigerant has its own specific requirements for recovery and evacuation, which must be understood beforehand. You can check the nameplate on the approach to identify the refrigerant used.

Here are **some additional tips for recovering refrigerants** quickly and efficiently:

- Use recovery equipment that is the correct size for the system being serviced.
- Use short, wide hoses to minimize pressure drop.
- Keep the ambient temperature as warm as possible.
- Repair any leaks in the system before beginning recovery.
- Evacuate the system to a low pressure before beginning recovery.
- Recover the refrigerant in a controlled manner, avoiding rapid pressure changes.

## (H) DEHYDRATION EVACUATION

Dehydration evacuation serves to eliminate water and water vapor from a refrigeration system.

If moisture is present in a refrigeration system that is in operation, it can produce highly corrosive and hazardous acids that can harm the system and pose a safety risk.

For optimal moisture removal from a system, evacuating it to a high vacuum level is advised. The necessary vacuum level may vary depending on the refrigerant type, but typically, it is around 500 microns or lower.

The factors that affect the speed and efficiency of evacuation are:

- Size of equipment being evacuated: The larger the equipment, the longer it will take to evacuate.

- Ambient temperature: The warmer the temperature, the faster it will evacuate. You may heat the refrigeration system to decrease the evacuation time.

- Amount of moisture in the system: The more moisture in the system, the longer it will take to evacuate.

- Size (capacity) of vacuum pump and suction line: For efficient evacuation, use a high-capacity vacuum pump with short, wide piping connections. Use hoses equal to or larger than the pump intake for optimal performance.

- Location of the vacuum gauge: The vacuum gauge (Micron Gauge) should be located as far as possible from the vacuum pump for the most accurate readings during evacuation.

Dehydration is considered complete once the vacuum gauge indicates that the system has achieved and maintained the required finished vacuum. It is impossible to over-evacuate a system.

Here are some additional tips for dehydration evacuation:

- Use a high-quality vacuum pump.
- Use the correct size vacuum pump for the size of the system being evacuated.
- Use short, wide vacuum hoses.
- Heat the refrigeration system to decrease the evacuation time.
- Monitor the vacuum gauge closely during evacuation.
- Evacuate the system to the required finished vacuum.

By following these tips, technicians can ensure that dehydration evacuation is performed safely and efficiently

# (I) SAFETY

When working with refrigerants, taking steps to protect yourself from potential hazards is essential. These hazards include:

- **Oxygen deprivation**: Refrigerants are heavier than air and can displace oxygen, causing a hypoxic environment. This can lead to dizziness, unconsciousness, or death. Safety precautions are crucial when working with refrigerants.

- **Cardiac effects**: Inhaling refrigerant vapors can irritate the respiratory tract and cause heart irregularities.

- **Frostbite**: Refrigerant can cause frostbite if it comes into contact with skin.

- **Long-term health effects**: Over time, being exposed to refrigerants can result in various negative health impacts, such as liver and kidney damage, as well as an increased risk of cancer.

To protect yourself from these hazards, you should always take the following safety precautions:

- Wear personal protective equipment (PPE), including safety glasses, gloves, and respirators.

- Use the proper tools and equipment: Make sure you have the right tools and equipment for the job, such as recovery cylinders, vacuum pumps, and leak detectors.

- Follow all safety procedures: Be sure to follow all safety procedures outlined in the manufacturer's instructions and the EPA's regulations.

**Specific safety precautions for working with refrigerants:**

- Never mix refrigerants: Mixing refrigerants can create hazardous compounds.

- Never fill a recovery cylinder more than 80% complete: This can create a safety hazard if the cylinder ruptures.

- Never use oxygen or compressed air to leak. Check a system: This can create an explosion hazard.

- Never heat a refrigerant cylinder with an open flame: This can create a fire hazard.

- Never cut or braze refrigerant lines on a charged unit: This can create a release of refrigerant.

- In the event of a significant release of refrigerant in a confined area, use a Self-Contained Breathing Apparatus (SCBA). This will protect you from inhaling the refrigerant vapors.

- If a large refrigerant leak occurs in an enclosed area, and SCBA is unavailable, immediately vacate and ventilate the area. This will help prevent oxygen deprivation.

- Never expose R-12 or R-22 to open flames or glowing hot metal surfaces: This can create hazardous compounds.

- Always review the material safety data sheets (MSDS): This will provide information on the specific hazards associated with the refrigerants you are working with.

By following these safety precautions, you can help protect yourself from the potential hazards of working with refrigerants.

By following these safety precautions, you can help to protect yourself from the potential hazards of working with refrigerants.

| ASHRAE Classification | Toxicity | Flammability |
|---|---|---|
| A1 | No toxicity | No flame propagation |
| A2 | Slightly toxic | No flame propagation |
| A3 | Moderately toxic | No flame propagation |
| B1 | Slightly toxic | Slightly flammable |
| B2 | Moderately toxic | Slightly flammable |
| B3 | Highly toxic | Highly flammable |

## (J) SHIPPING

To transport refrigerants safely, use DOT (Department of Transportation) - approved cylinders designed to withstand pressure and prevent leaks. Label the cylinders properly with information such as the type and quantity of refrigerant and the shipper's name and address, as required by the EPA.

Regulations set by the DOT govern the shipping of refrigerants. These regulations encompass various areas, including:

- The type of cylinder that can be used to ship refrigerants
- The labeling requirements for refrigerant cylinders
- The packaging requirements for refrigerant cylinders
- The shipping documentation requirements for refrigerant cylinders
- The training requirements for individuals who ship refrigerant cylinders

To guarantee the secure shipment of refrigerants, it is essential to adhere to the regulations set out by the EPA and DOT.

Here are some additional tips for shipping refrigerants:

- Never ship refrigerants in the same container as other hazardous materials.

- Please always look over the cylinders for leaks before shipping them.

- Never overload the cylinders.

- Please make sure to store the cylinders in a cool, dry place.

- Never expose the cylinders to heat or open flames.

- Never ship refrigerants to a location that is not authorized to receive them.

For safe shipping of refrigerants, follow these tips and check DOT regulations for more info.

- The type of cylinder that can be used to ship refrigerants: It is essential to ensure that DOT-approved cylinders are used when shipping refrigerants. Look for the letter "DOT" followed by a four-digit number to ensure compliance.

- The labeling requirements for refrigerant cylinders: Refrigerant cylinders must be labeled with the following information:

1. The type of refrigerant
2. The quantity of refrigerant
3. The name and address of the shipper
4. The hazard class of the refrigerant
5. The United Nations identification number for the refrigerant

- Packaging requirements for refrigerant cylinders: To avoid leaks, it's essential to package refrigerant cylinders properly. This might mean using extra materials like shrink wrap or foam to ensure safe transportation.

- The shipping documentation requirements for refrigerant cylinders: The shipper must provide a manifest with contents, shipper and consignee info, and refrigerant hazard class to the carrier.

- The training requirements for individuals who ship refrigerant cylinders: Proper training is necessary when transporting refrigerant cylinders, and a certified instructor should provide it.

By following the DOT's regulations, you can help ensure refrigerants' safe shipping.

### R-410A

R-410A is a blend of HFC-32 and HFC-125 used in modern ACs and heat pumps. It's eco-friendly and efficient but harder to handle than older refrigerants like R-22.

EPA 608 certification is required for technicians who work on air conditioners and heat pumps containing refrigerant. This certification ensures the safe handling and servicing of refrigerant systems.

EPA 608 certification covers all refrigerants, including R-410A. However, additional training is recommended for technicians working with this specialized refrigerant.

Technicians often pursue further training despite EPA 608 certification. This training covers R-410A's higher pressure, specialized tools, and charging/recovery/evacuation best practices for safe handling.

# 3. Type I (Small Appliances) of EPA 608 Certification

# 3. TYPE 1 (SMALL APPLIANCES) SECTION OF EPA 608 CERTIFICATION

This chapter focuses on the scope of the Type 1 section and its explanation.

## Scope of Type I Section (Syllabus for the Type I Exam)
*(source: https://www.epa.gov/section608)*

**(1) Recovery Requirements**

- Definition of "small appliance"

- Evacuation requirements for small appliances with and without working compressors using recovery equipment manufactured before November 15, 1993

- Evacuation requirements for small appliances with and without working compressors using recovery equipment manufactured after November 15, 1993

**(2) Recovery Techniques**

- Use of pressure and temperature to identify refrigerants and detect noncondensable

- Methods to recover refrigerant from small appliances with inoperative compressors using a system-dependent or "passive" recovery device (e.g., heat and sharply strike the compressor, use a vacuum pump with a non-pressurized recovery container)

- Need to install both high and low side access valves when recovering refrigerant from small appliances with inoperative compressors

- Need to operate operative compressors when recovering refrigerant with a system-dependent ("passive") recovery device

- Should remove solderless access fittings after service

- Hydrofluorocarbon (HFC)-134a (also called R-134a) as likely a substitute for chlorofluorocarbon (CFC)-12 (also called R-12)

**Safety**

- Decomposition products of refrigerants at high temperatures

### Type 1

- Installation, maintenance, repair, and disposal of small appliances that contain refrigerant

- Small appliances are defined as those that contain less than 5 pounds of refrigerant and operate at pressures less than 150 psig

- Examples of small appliances include refrigerators, freezers, window air conditioners, and dehumidifiers

# EXAM-SPECIFIC POINTS FOR THE TYPE I SECTION

## (A) RECOVERY REQUIREMENTS

To work with small appliances, including refrigerant recovery, you need EPA Type I Certification. Only certified technicians can sell refrigerants for servicing or installing refrigeration and air conditioning equipment.

**Definition of "small appliance":**

As the EPA defines, a small appliance is a product that is wholly manufactured, charged, and sealed at the factory and contains five pounds or less of refrigerant. Some examples of small appliances include:

- Refrigerators and freezers designed for home use

- Room air conditioners (including window air conditioners and packaged terminal air conditioners)

- Packaged terminal heat pumps

- Dehumidifiers

- Under-the-counter ice makers

- Vending machines

- Drinking water coolers

**Small Appliance Certification Requirements**

Individuals who work with refrigerants while servicing, maintaining, or repairing small appliances must hold a Type I or Universal certification. The sale of CFCs and HCFCs is limited to technicians who have been certified.

## MVAC Systems

Motorized vehicle air conditioning (MVAC) systems are not considered small appliances and require a separate certification.

## Refrigerant Recovery Requirements

When technicians service a small appliance, they must recover all refrigerant. This refrigerant must be recovered and stored in an approved container, such as a recovery cylinder. It is important to note that the refrigerant cannot be released into the atmosphere.

Technicians who work on small appliances must abide by the strict regulations set by the EPA regarding the handling and recovery of refrigerants. Certification is necessary for such professionals, and they must adhere to all relevant rules. Here are some extra tips for technicians working on small appliances.:

- Always wear personal protective equipment (PPE) when working with refrigerants.

- Never vent refrigerant to the atmosphere.

- Recover all refrigerant from the system before servicing.

- Dispose of refrigerant properly.

By following these tips, technicians can help to protect the environment and ensure the safe handling of refrigerants.

# Evacuation requirements for small appliances with and without working compressors using recovery equipment

(manufactured before and after November 15, 1993)
**Appliances with a charge under 5 pounds**

(A) Using a recovery device manufactured **before November 15, 1993**

- Compressor running or not, 80% of charge or 4 inches Hg vacuum.

- Compressor running 90% or 4 inches Hg vacuum.

(B) Using a recovery device manufactured on or **after November 15, 1993**:

- The compressor is not running, 80% of charge or 4 inches Hg vacuum

- Compressor running, 90% of charge or 4 inches Hg vacuum

**Fittings**: Low-loss fittings required

**Approvals:**

- Before November 15, 1993: None

- After November 15, 1993: EPA Lab Approved

| Recovery Device Manufacture Date | Compressor Running | Compressor Not Running | Fittings | Approvals |
|---|---|---|---|---|
| Before November 15, 1993 | 80% of charge or 4 inches Hg vacuum | 80% of charge or 4 inches Hg vacuum | Low-loss required | None |
| After November 15, 1993 | 90% of charge or 4 inches Hg vacuum | 80% of charge or 4 inches Hg vacuum | Low-loss required | EPA Lab Approved |

# (B) RECOVERY TECHNIQUES

## Recovery Techniques

The process of refrigerant recovery involves the removal of refrigerant from a system to ensure its proper recycling or disposal.

Recovering refrigerant from small appliances is vital in preventing environmental harm and safeguarding the well-being of technicians.

There are two main methods for recovering refrigerant from small appliances:

- **Active Recovery:** The most efficient method is active recovery, which involves using a recovery machine to recover refrigerant from a system. However, this method requires special equipment.

- **Passive Recovery**: Passive recovery method involves using a non-pressurized container to collect the refrigerant released from the system. While it may not be as efficient as active recovery, it does not require any special equipment.

## Use of Pressure and Temperature to Identify Refrigerants and Detect Noncondensables

By analyzing the pressure and temperature of a refrigerant, it is possible to determine its type and detect the presence of noncondensable gases.

These gases, which do not condense under normal refrigeration temperatures, can infiltrate the system through leaks or improper recovery equipment.

Unfortunately, noncondensable can diminish the efficiency of the system and even harm compressors.

# Recovery Techniques

The process of refrigerant recovery involves the removal of refrigerant from a system to ensure its proper recycling or disposal.

Recovering refrigerant from small appliances is vital in preventing environmental harm and safeguarding the well-being of technicians.

There are two main methods for recovering refrigerant from small appliances:

- **Active Recovery:** The most efficient method is active recovery, which involves using a recovery machine to recover refrigerant from a system. However, this method requires special equipment.

- **Passive Recovery**: Passive recovery method involves using a non-pressurized container to collect the refrigerant released from the system. While it may not be as efficient as active recovery, it does not require any special equipment.

## Use of Pressure and Temperature to Identify Refrigerants and Detect Noncondensables

By analyzing the pressure and temperature of a refrigerant, it is possible to determine its type and detect the presence of noncondensable gases.

These gases, which do not condense under normal refrigeration temperatures, can infiltrate the system through leaks or improper recovery equipment.

Unfortunately, noncondensable can diminish the efficiency of the system and even harm compressors.

**Identifying refrigerants:** Use a pressure/temperature chart to identify the refrigerant in a system. Compare the system's pressure and temperature to the chart to determine the refrigerant type.

- **Detecting noncondensable**: Noncondensable gases, which don't condense at refrigeration temperatures, can harm compressors and reduce system efficiency if they enter due to leaks or improper equipment use.

To check for noncondensable in a system, use a pressure/temperature chart. If values are outside the listed range, noncondensable's are likely present.

Other methods include using a noncondensable gas detector or vacuum pump. Removing noncondensable is important for system efficiency and compressor safety.

| Refrigerant | Pressure (psi) |
|---|---|
| R-134a | 100 |
| R-22 | 150 |
| R-410A | 225 |
| R-404A | 275 |
| R-507 | 300 |

Use this table to identify the type of refrigerant in a system. Match the system's pressure to the values listed. For example, if the pressure is 100 psi at 40°F, the refrigerant is likely R-134a.

Here is an example of how to use the table to identify the type of refrigerant in a system:

- Measure the system pressure at 40 degrees Fahrenheit.

- Compare the system pressure to the values listed in the table.

- The refrigerant in the system is the one that has a pressure that is closest to the measured pressure.

- **Detecting noncondensable**: Noncondensable gases, which don't condense at refrigeration temperatures, can harm compressors and reduce system efficiency if they enter due to leaks or improper equipment use.

To check for noncondensable in a system, use a pressure/temperature chart. If values are outside the listed range, noncondensable's are likely present.

Other methods include using a noncondensable gas detector or vacuum pump. Removing noncondensable is important for system efficiency and compressor safety.

Refrigerants can be identified with a refrigerant identifier or by sending a sample to a lab. Proper identification is essential for safe handling and disposal.

Here is another example of how to use pressure and temperature to detect noncondensable:

- Measure the system pressure and temperature.

- Compare the system pressure and temperature to the values listed in a pressure/temperature chart for the type of refrigerant that is in the system.

- If the system pressure and temperature are not within the range listed on the chart, they likely are noncondensable in the system.

Removing noncondensable from refrigeration systems is essential for efficient and damage-free compressor operation. A vacuum pump can be used to create a vacuum and eliminate these noncondensable.

## Methods to Recover Refrigerant from Small Appliances with Inoperative Compressors using a System-Dependent or "Passive" Recovery Device (e.g., Heat and Sharply Strike the Compressor, Use a Vacuum Pump with a Non-pressurized Recovery Container)

If the compressor of a small appliance is not functioning, a passive recovery method may be required. There are various ways to aid passive recovery:

- **Use the appliance compressor to pump refrigerant:** If the compressor is still capable of running, it can be used to pump refrigerant into the non-pressurized container.

- **Use the pressure of the refrigerant to facilitate transfer:** The pressure of the refrigerant can be used to force it into the non-pressurized container.

- **Heat the compressor and strike it:** Heating it and striking it with a mallet can help release refrigerant from the compressor oil.

- **Use a vacuum pump**: A vacuum pump can be used to remove noncondensable from the system, which can help improve passive recovery efficiency.

## Need to Install Both High and Low Side Access Valves when Recovering Refrigerant from Small Appliances with Inoperative Compressors

When the compressor in a small appliance is not working, installing access valves on both the high and low sides is necessary for refrigerant recovery. This step ensures that all refrigerant is completely removed from the system.

## Need to Operate Operative Compressors when Recovering Refrigerant with a System-Dependent ("Passive") Recovery Device

When using a system-dependent ("passive") recovery device to recover refrigerant from a small appliance, it is important to operate the compressor if it is functional. This will help to remove the refrigerant from the system efficiently.

## Should Remove Solderless Access Fittings at Conclusion of Service

It is recommended to remove solderless access fittings from small appliances after service as they have a tendency to leak over time.

**Hydrofluorocarbon (HFC)-134a (also called R-134a) as Likely Substitute for Chlorofluorocarbon (CFC)-12 (also called R-12)**

When it comes to replacing CFC-12 in small appliances, HFC-134a is a viable option. This refrigerant is ozone-friendly and possesses properties that are similar to CFC-12.

## (C) SAFETY

### Decomposition products of refrigerants at high temperatures

When refrigerants are subjected to high temperatures, they may break down into dangerous substances. The particular substances produced during decomposition vary depending on the refrigerant type, but they may include:

- **Phosgene gas:** Phosgene gas is extremely toxic and can lead to fatalities. It is a colorless and odorless gas, making it easy to inhale. Exposure to phosgene gas can result in respiratory issues, skin burns, and even death.

- **Hydrochloric acid**: Hydrochloric acid is a highly corrosive acid that can potentially cause serious burns to both the skin and eyes. This colorless gas has a strong, pungent odor and can also cause damage to metal and various other materials.

- **Hydrofluoric acid:** Be cautious when handling it, as it is a highly corrosive substance that can cause severe burns to the skin and eyes. This colorless gas gives off a sweet odor but can also cause damage to metals and other materials.

Refrigerants can pose a serious hazard when they decompose at high temperatures. It is crucial to take precautionary measures to prevent refrigerants from being exposed to high temperatures.

This can be achieved by utilizing appropriate equipment and procedures while working with refrigerants.

Here are some additional safety tips to keep in mind when working with refrigerants:

- Wearing the appropriate personal protective equipment (PPE) is essential when working with refrigerants. This includes safety glasses, gloves, and a long-sleeved shirt and pants.

- Never use a flame to heat any component containing refrigerant.

- Always use a pressure regulator when using nitrogen.

- If refrigerant is significantly released, evacuate and ventilate the area.

- It is essential only to use recovery or recycling machines expressly certified for the refrigerant being used. Never attempt to recover flammable refrigerants in a device not certified for that purpose.

- Always ensure that the refrigeration system, recovery unit, and recovery cylinder are all adequately grounded.

Following these safety tips can help prevent accidents and injuries when working with refrigerants.

# 4. Type 2 (High-Pressure) of EPA 608 Certification

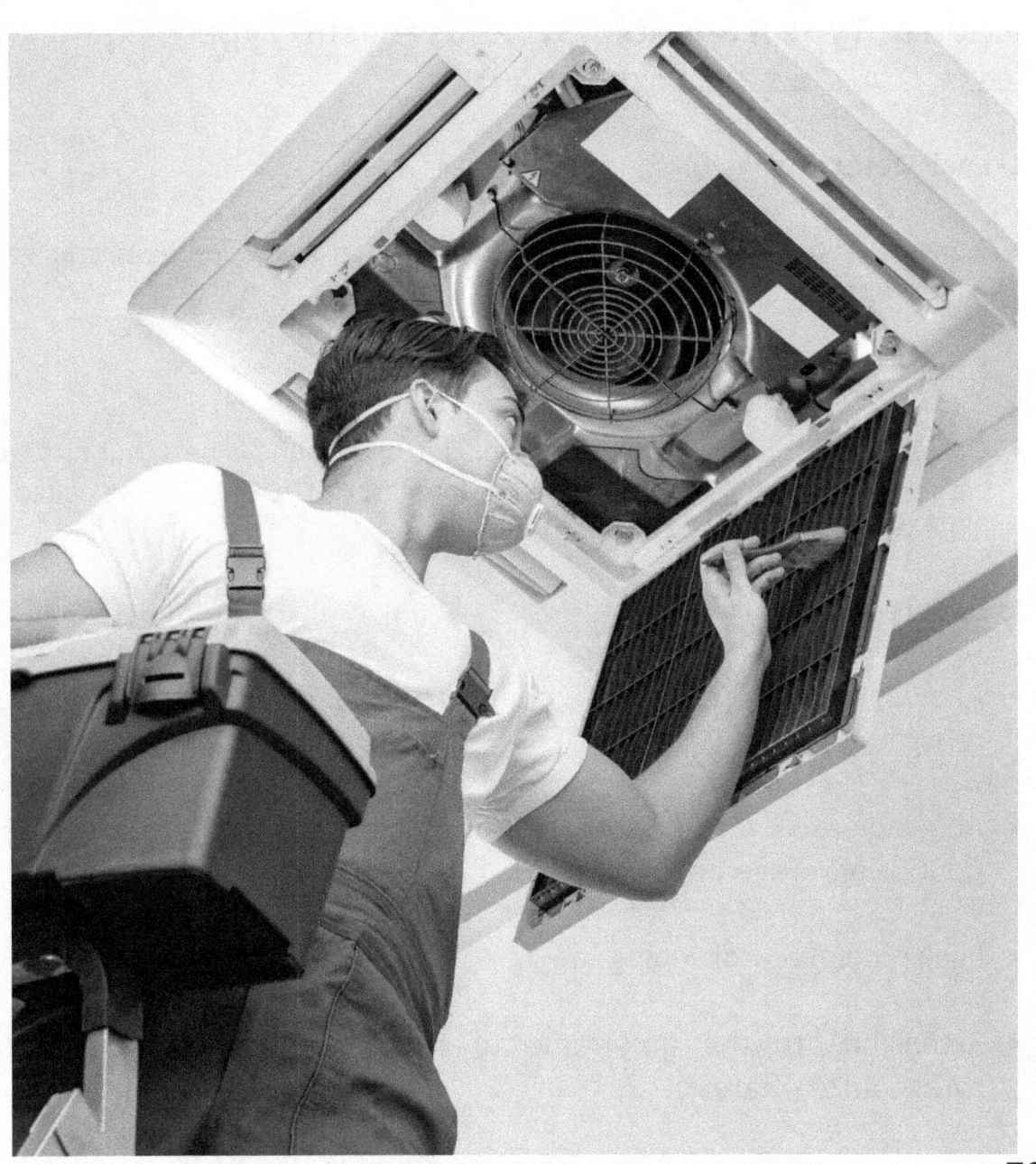

# 4. TYPE 2 (HIGH-PRESSURE) SECTION OF EPA 608 CERTIFICATION

This chapter focuses on the scope of the Type 2 section and its explanation.

## Scope of Type II Section (Syllabus for the Type 2 Exam)
*(source: https://www.epa.gov/section608)*

**(1) Leak Detection**

- Signs of leakage in high-pressure systems (excessive superheat, traces of oil for hermetics)

- Need to leak test before charging or recharging equipment

- Order of preference for leak test gases [nitrogen alone is best, but nitrogen with a trace quantity of hydrochlorofluorocarbon (HCFC)-22 (also called R-22) is better than pure refrigerant]

**(2) Leak repair requirements**

- Allowable leak rate for commercial and industrial process refrigeration

- Allowable leak rate for other appliances containing more than 50 pounds of refrigerant

- Leak repair recordkeeping

- Extensions to the timeframe to repair leaks that exceed the threshold leak rate

## (3) Recovery Techniques

- Recovering liquid at the beginning of the recovery process speeds up the process

- Other methods for speeding recovery (chilling recovery vessel, heating appliance, or vessel from which refrigerant is being recovered)

- Methods for reducing cross-contamination and emissions when a recovery or recycling machine is used with a new refrigerant

- Need to wait a few minutes after reaching the required recovery vacuum to see if system pressure rises (indicating that there is still liquid refrigerant in the system or the oil)

## (4) Recovery Requirements

- Evacuation requirements for high-pressure appliances in each of the following situations:

1. Disposal
2. Major versus non-major repairs
3. Leaky versus non-leaky appliances
4. Appliance (or component) containing less versus more than 200 pounds
5. Recovery/recycling equipment built before versus after November 15, 1993

- Definition of "major" repairs

- Prohibition on using system-dependent recovery equipment on systems containing more than 15 pounds of refrigerant

## (5) Refrigeration

- How to identify refrigerant in appliances

- Pressure-temperature relationships of common high-pressure refrigerants [may use standard temperature-pressure chart--be aware of the need to add 14.7 to translate pounds per square inch gauge (psig) to pounds per square inch absolute (psia)

- Components of high-pressure appliances (receiver, evaporator, accumulator, etc.) and state of refrigerant (vapor versus liquid) in them

- The idea that hydrocarbons are not approved for retrofits

## (6) Safety

- Shouldn't energize hermetic compressors under vacuum

- Equipment room requirements under American Society of Heating, Refrigerating, and Air-Conditioning Engineers (ASHRAE) Standard 15 (oxygen deprivation sensor with all refrigerants)

# EXAM-SPECIFIC POINTS FOR THE TYPE 2 SECTION

## (A) LEAK DETECTION

### Signs of Leakage in High-Pressure Systems

Several signs can indicate a leak in a high-pressure system. Some of the most common signs include:

- **Low refrigerant charge**: This can be caused by a leak or other problems. There is likely a leak if the system cannot maintain proper pressure.

- **Excessive superheat** occurs when the refrigerant cannot absorb enough heat from the evaporator. A low refrigerant charge or a leak can cause this.

- **Traces of oil**: Oil can leak out of the compressor or other components in the system. If you see oil around the system, it is a sign of a leak.

- **Visible evidence of a leak** may include bubbles, frost, or other signs of refrigerant escaping. You must check the system immediately if you see any visual leak evidence.

### Need to Leak Test Before Charging or Recharging Equipment

Before charging or recharging a high-pressure system with refrigerant, it is crucial to perform a leak test to ensure no leaks in the system.

This test will help confirm that the system is leak-free before adding refrigerant.

There are several different ways to leak-test a high-pressure system. Some of the most common methods include:

- Pressurizing the system with nitrogen is an excellent way to find large leaks.

- Using an electronic leak detector is a good way to find small leaks.

- Using a halide torch is also a way to find leaks in hard-to-reach areas.

After conducting a leak test and fixing any leaks, the system can be charged or recharged with refrigerant. This ensures the safety of the system.

**Order of Preference for Leak Test Gases**

- **Nitrogen alone**
- **Nitrogen with a trace quantity of R-22**
- **Other gases**

When testing for leaks in high-pressure systems, nitrogen is the best gas to use. Nitrogen is an inert gas that won't react with refrigerants, making it safe for all methods.

A small amount of R-22 can be used as a leak test gas if nitrogen is unavailable. While R-22 is a refrigerant, it can be used safely in small quantities. It is essential to avoid using other gases for leak testing, such as oxygen or carbon dioxide. These gases can react with refrigerants and create dangerous conditions.

**Allowable Annual Leak Rate**

It's important to adhere to specific leak detection guidelines to maintain the safety and efficiency of high-pressure systems. The allowable annual leak rate can vary based on the system type and refrigerant quantity.

For instance, commercial and industrial process refrigeration equipment containing over 50 pounds of refrigerant should be fixed if the leak surpasses 35% of the annual charge.

All other equipment with more than 50 pounds of refrigerant must be repaired if the leak rate exceeds 15% of the monthly charge. By following these guidelines, you can help guarantee the safety and efficiency of your high-pressure systems.

# (B) LEAK REPAIR REQUIREMENTS

## Allowable Leak Rate

- **Commercial and Industrial Process Refrigeration:** An appliance with over 50 pounds of refrigerant must be repaired if the annual leak rate exceeds 35% of the charge.

- **Other Appliances Containing More Than 50 Pounds of Refrigerant:** If an appliance holds over 50 pounds of refrigerant and is not used for commercial or industrial process refrigeration, it must be fixed when the yearly leak rate exceeds 15% of the charge.

## Leak Repair Recordkeeping

It is necessary to keep records of leak inspections, initial leak verification, and follow-up verification tests for a period of three (3) years.

The records should contain the following details:

- Date of the leak inspection

- Name of the technician who performed the inspection

- Results of the inspection

- Date of the initial leak verification test

- Results of the initial leak verification test

- Date of the follow-up verification test

- Results of the follow-up verification test

**Extensions to the Timeframe to Repair Leaks That Exceed the Threshold Leak Rate**

In certain circumstances, repairing a leak may take longer than expected.

For example, cold storage facilities may be allowed to continue using a faulty appliance with a leak for up to 12 months before it needs to be repaired or replaced.

To qualify for an extension on the repair timeline, the following requirements must be met:

- The owner or operator of the appliance must submit a written request to the EPA.

- The request must include the following information:

1. The name and address of the owner or operator of the appliance
2. The type and size of the appliance
3. The date the leak was detected
4. The reason for the request for an extension

- The EPA will review the request and may grant an extension for up to 12 months.

## (C) RECOVERY TECHNIQUES

**Recovering Liquid at the Beginning of Recovery Speeds Up the Process**

- Extracting as much liquid refrigerant as possible is crucial when starting the refrigerant recovery. This will accelerate the process and reduce the amount of refrigerant that escapes into the atmosphere.

- There are a few methods to recover liquid refrigerant. One way is to connect the recovery hoses to the appliance's liquid line. Another option is to connect the recovery hoses to the condenser outlet if it's situated below the receiver.

- Once the liquid refrigerant has been recovered, it is possible also to recover any remaining vapor.

## Other Methods for Speeding Recovery

- Besides recovering the liquid refrigerant at the start of the process, a few other techniques are available to expedite the recovery process.

- One method is to cool the recovery cylinder. This will help lower the pressure in the cylinder, allowing the refrigerant to flow more easily.

- Another method is to heat the appliance. This will help increase the pressure in the appliance, allowing the refrigerant to flow more easily.

- It is essential to use caution when heating the appliance, as too much heat can damage it.

## Methods for Reducing Cross-Contamination and Emissions

- When recovering the refrigerant, it is essential to prevent cross-contamination and emissions.

- One way to prevent cross-contamination is to use separate recovery hoses for each type of refrigerant.

- When working with refrigerants, it is advisable to purge the recovery equipment before switching to a different type to prevent cross-contamination.

- Before transferring refrigerant to the cylinder, it is crucial to use a vacuum pump to evacuate the recovery cylinder. This process eliminates any non-condensable gases from the cylinder, which can cause emissions.

## Need to Wait a Few Minutes After Reaching Required Recovery Vacuum to See if System Pressure Rises

- Waiting a few minutes after achieving the necessary recovery vacuum before disconnecting the recovery hoses is essential. This allows sufficient time for any residual liquid refrigerant to vaporize.

- If the pressure in the system increases after a few minutes, it suggests that there is still refrigerant in liquid form present in the system. In such a situation, it is recommended to continue the recovery process until the pressure in the system stabilizes.

## (D) RECOVERY REQUIREMENTS

Evacuation requirements for high-pressure appliances in each of the following situations:

### Evacuation Requirements for Disposal

- When removing a high-pressure appliance, you must recover the refrigerant and then evacuate the appliance to atmospheric pressure (0 psig).

- This applies to all appliances, regardless of the refrigerant charge size or whether the appliance has a leak.

### Evacuation Requirements for Major Repairs

When conducting a significant repair on a high-pressure device, it is necessary to recover the refrigerant and evacuate the appliance to a vacuum of either 0 or 10 inches Hg, depending on the appliance's age and the type of refrigerant used.

A major repair is defined as any maintenance, service, or repair that involves the removal of any of the following components:
- The compressor
- The condenser
- The evaporator
- An auxiliary heat exchanger coil

**Evacuation Requirements for Leaky Appliances**

- If a high-pressure appliance has a leak preventing it from being evacuated to the required level, it must be evacuated to atmospheric pressure (0 psig) before opening.

- This is to prevent the release of refrigerant into the atmosphere.

**Evacuation Requirements for Appliances Containing Less Than or More Than 200 Pounds of Refrigerant**

- The evacuation requirements for high-pressure appliances vary depending on the refrigerant charge size.

- Appliances with less than 200 pounds of refrigerant must be vacuumed to either 0 inches Hg or 10 inches Hg, depending on the refrigerant type and appliance age.

- Appliances with over 200 pounds of refrigerant must be vacuumed to 4 or 15 inches Hg, based on refrigerant type and appliance age.

**Evacuation Requirements for Recovery/Recycling Equipment Manufactured Before or After November 15, 1993**

- Any recovery and recycling equipment produced or imported after November 15, 1993, must be certified by an EPA-approved and third-party laboratory as per ARI 740.

- This ensures that the equipment is capable of recovering refrigerant to the required levels.

**Definition of "Major Repairs"**

A significant repair involves removing any of these components: maintenance, service, or repair.

- The compressor

- The condenser

- The evaporator

- An auxiliary heat exchanger coil

**Prohibition on Using System-Dependent Recovery Equipment on Systems Containing More Than 15 Pounds of Refrigerant**

- Refrigerant appliances over 15 pounds cannot use system-dependent recovery equipment (passive recovery equipment).

- This is because system-dependent recovery equipment cannot recover refrigerant to the required levels.

## (E) REFRIGERATION

### How to Identify Refrigerant in Appliances

- The type and amount of refrigerant used in an appliance can be identified by looking at the nameplate.

- To find out what refrigerant your appliance uses, consult a pressure-temperature chart. This chart relates pressure and temperature to different refrigerants, allowing you to determine the exact type used by measuring the pressure and temperature in your appliance.

### Pressure-Temperature Relationships of Common High-Pressure Refrigerants

- The temperature of a refrigerant is directly proportional to its pressure: as pressure increases, temperature increases, and as pressure decreases, temperature decreases.

- Refrigerants' pressure and temperature relationships can vary depending on the type and conditions of use.

- The following table shows the pressure-temperature relationships of some common high-pressure refrigerants.

| Refrigerant | Pressure (psig) | Temperature (°F) |
|:---:|:---:|:---:|
| R-22 | 70 | 40 |
| R-134a | 70 | 32 |
| R-410A | 250 | 104 |
| R-404A | 250 | 86 |

## Components of High-Pressure Appliances (Receiver, Evaporator, Accumulator, etc.) and State of Refrigerant (Vapor versus Liquid) in Them

The following are the components of a high-pressure refrigeration system:
Compressor: The compressor is responsible for compressing the refrigerant gas.

**Condenser**: The condenser condenses the refrigerant gas into a liquid.

**Receiver**: The receiver is responsible for storing the liquid refrigerant.

**Expansion device**: The expansion device converts the liquid refrigerant into a vapor.

**Evaporator:** The evaporator evaporates the refrigerant vapor, which absorbs heat from the surrounding air or water.

**Accumulator**: The accumulator collects liquid refrigerant that may condense in the evaporator.

The state of the refrigerant in each of these components depends on the pressure and temperature of the refrigerant.

- In the compressor, the refrigerant is a gas.
- In the condenser, the refrigerant is a liquid.
- In the receiver, the refrigerant is a liquid.
- In the expansion device, the refrigerant is a liquid-vapor mixture.
- In the evaporator, the refrigerant is a vapor.
- The refrigerant may be a liquid or a liquid-vapor mixture in the accumulator.

## The idea that Hydrocarbons are Not Approved for Retrofits

- Appliances designed initially for certain refrigerants cannot be retrofitted with hydrocarbon refrigerants.
- Hydrocarbons are flammable and may cause fire hazards if not handled correctly.
- If you're planning to use a hydrocarbon refrigerant in an appliance, it's essential to consult with the manufacturer first to ensure compatibility.

# (F) SAFETY

## Shouldn't Energize Hermetic Compressors Under Vacuum

- Starting a compressor under a vacuum can damage it due to low pressure, causing refrigerant flow issues.

- Open the discharge service valve first if you need to start a compressor under a vacuum. This will help lubricate the compressor and prevent damage.

## Equipment Room Requirements under ASHRAE Standard 15

- Equipment rooms with refrigerants must have a refrigerant sensor, alarm, and ventilation under ASHRAE Standard 15.

- The refrigerant sensor will detect the presence of refrigerant in the air and sound an alarm if the concentration reaches a certain level.

- The ventilation system will then be activated to remove the refrigerant from the air.

- This helps prevent refrigerant buildup in the air, which can harm people and animals.

## Oxygen Deprivation Sensor with All Refrigerants

- ASHRAE Standard 15 also requires equipment rooms containing refrigerants to be equipped with an oxygen deprivation sensor.

- This sensor will detect the oxygen level in the air and sound an alarm if the level drops below a certain level.

- This helps to prevent people and animals from being exposed to low levels of oxygen, which can be harmful or even fatal.

| Safety Requirement | Description |
| --- | --- |
| Do not energize hermetic compressors under vacuum | Starting a compressor under vacuum will damage the compressor. |
| Equipment rooms that contain refrigerants must be equipped with a refrigerant sensor, an alarm, and a ventilation system | The refrigerant sensor will detect the presence of refrigerant in the air and sound an alarm if the concentration reaches a certain level. The ventilation system will then be activated to remove the refrigerant from the air. This helps to prevent the buildup of refrigerant in the air, which can be harmful to people and animals. |
| Equipment rooms that contain refrigerants must also be equipped with an oxygen deprivation sensor | This sensor will detect the level of oxygen in the air and sound an alarm if the level drops below a certain level. This helps to prevent people and animals from being exposed to low levels of oxygen, which can be harmful or even fatal. |

# 5. Type 3 (Low-pressure) of EPA 608 Certification

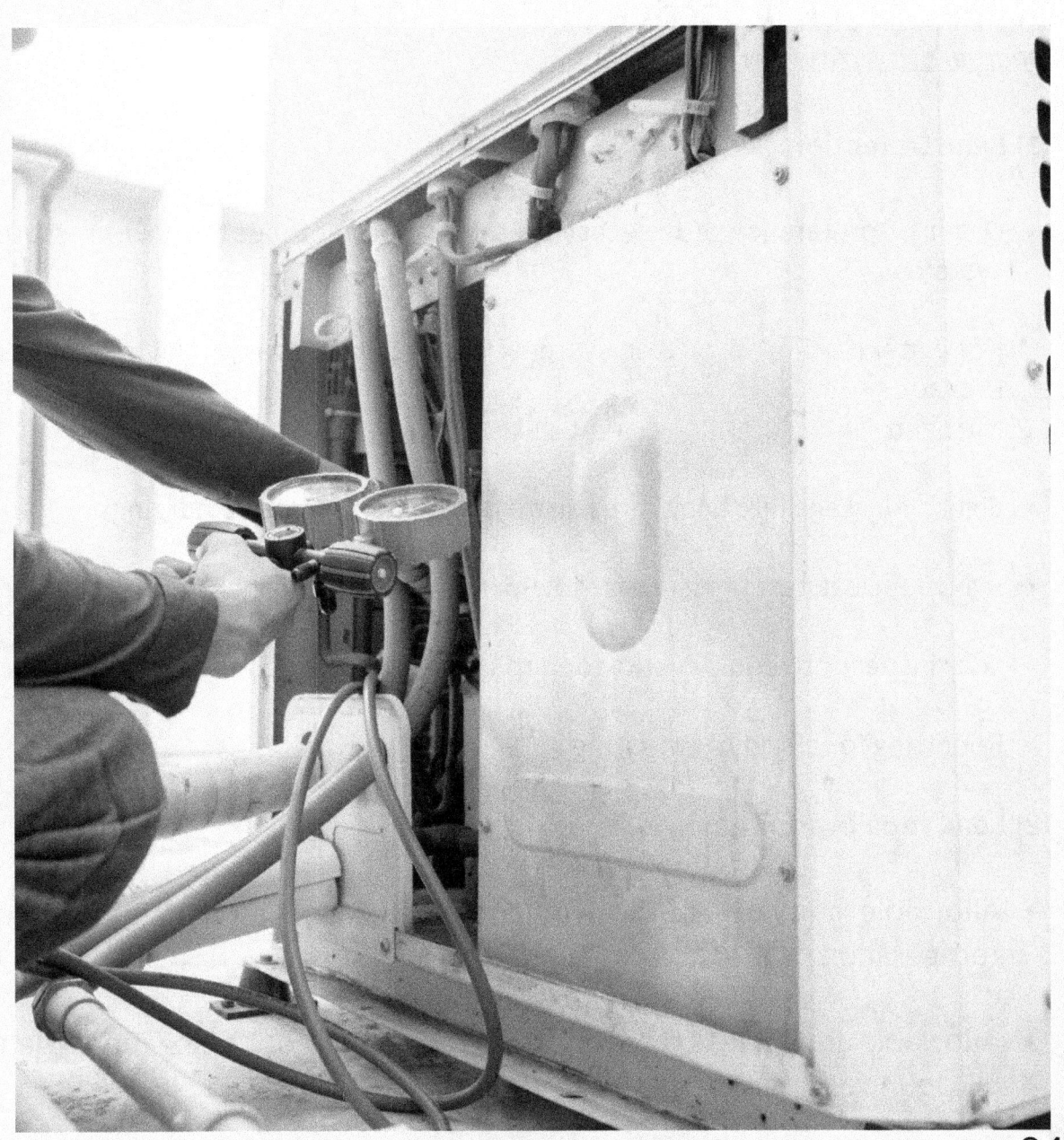

# 5. TYPE 3 (LOW-PRESSURE) SECTION OF EPA 608 CERTIFICATION

This chapter focuses on the scope of the Type 3 section and its explanation.

**Scope of Type III Section** (Syllabus for the Type 3 Exam)
*(source: https://www.epa.gov/section608)*

### (1) Leak Detection

- Order of preference of leak test pressurization methods for low-pressure systems

  1. Hot water method or built-in system heating/pressurization device such as Prevac
  2. Nitrogen

- Signs of leakage into a low-pressure system (e.g., excessive purging)

- Maximum leak test pressure for low-pressure centrifugal chillers

- Leak inspection requirements for appliances that exceed the leak rate

- Reporting for chronically leaking appliances

### (2) Leak repair requirements

- Allowable annual leak rate for commercial and industrial process refrigeration

- Allowable annual leak rate for other appliances containing more than 50 pounds of refrigerant

## (3) Recovery Techniques

- Recovering liquid at the beginning of the recovery process speeds up the process

- Need to recover vapor in addition to liquid

- Need to heat oil to 130°F before removing it to minimize the refrigerant release

- Need to circulate or remove water from chiller during refrigerant evacuation to prevent freezing

- High-pressure cut-out level of recovery devices used with low-pressure appliances

## (4) Recharging Techniques

- Need to introduce vapor before liquid to prevent water from freezing in the tubes.

- Need to charge centrifugal through evaporator charging valve.

## (5) Recovery Requirements

- Evacuation requirements for low-pressure appliances in each of the following situations:

1. Disposal
2. Major versus non-major repairs
3. Leaky versus non-leaky appliances
4. Appliance (or component) containing less versus more than 200 pounds
5. Recovery/recycling equipment built before versus after November 15, 1993

- Definitions of "major" and "non-major" repairs

- Allowable methods for pressurizing a low-pressure system for a non-major repair (controlled hot water and system heating/pressurization device such as Prevac)

- Need to wait a few minutes after reaching the required recovery vacuum to see if system pressure rises (indicating that there is still liquid refrigerant in the system or the oil)

## (6) Refrigeration

- Purpose of purge unit in low-pressure systems

- Pressure-temperature relationships of low-pressure refrigerants

## (7) Safety

- Equipment room requirements under ASHRAE Standard 15 (oxygen deprivation sensor with all refrigerants)

- Under ASHRAE Standard 15, need to have an equipment room refrigerant sensor for R-123

# EXAM-SPECIFIC POINTS FOR THE TYPE 3 SECTION

## (A) LEAK DETECTION

### Order of Preference of Leak Test Pressurization Methods for Low-Pressure Systems

- Low-pressure systems operate below atmospheric pressure, so any leaks in the gaskets or fittings will allow air and moisture to enter the system. This can lead to several problems, including:

1. Reduced system efficiency
2. Increased operating costs
3. Damage to the compressor or other components
4. Release of refrigerant into the atmosphere

- To prevent leaks in low-pressure systems, it's essential to conduct regular leak tests.

- Low-pressure systems can be leak-tested using various methods, with the hot water method or a built-in system heating/pressurization device being the most effective.

- If the available methods are not feasible, nitrogen can be used to test for leaks in the system. However, caution should be exercised to ensure the pressure does not exceed ten psig.

### (a) Hot Water Method or Built-In System Heating/Pressurization Device Such as Prevac

The most efficient way to leak-test a low-pressure system is by using hot water. Circulate hot water through the design and use a leak detector to check for leaks. Follow the manufacturer's instructions if the system has a built-in heating/pressurization device.

### (b) Nitrogen

If you can't use the hot water method or a built-in system heating/pressurization device to test for leaks, you can use nitrogen instead. Nitrogen is safe and won't react with the refrigerant or system components. Connect a nitrogen cylinder, pressurize the system, and use a leak detector. Remember to stay within ten psig.

### Signs of Leakage into a Low-Pressure System (e.g. Excessive Purging)

- Several signs can indicate a leak in a low-pressure system. These signs include:

1. Excessive purging of the purge unit.
2. High head pressure.
3. Moisture in the purge unit.
4. A rise in pressure during a standard vacuum test.

- If you notice any of the following signs, it is essential to inspect the system for leaks immediately.

### Maximum Leak Test Pressure for Low-Pressure Centrifugal Chillers

- The maximum test pressure for low-pressure centrifugal chillers is 10 psig. Pressures higher than this can damage the system.

- It is crucial to follow the instructions provided by the manufacturer when performing leak tests on low-pressure centrifugal chillers.

### Leak Inspection Requirements for Appliances that Exceed the Leak Rate

Appliances that exceed the leak rate require biannual inspections by a qualified technician.

The inspection must include the following steps:

- A visual inspection of the system for leaks.

- A pressure test of the system.
- A leak detection test of the system.

If any leaks are found, they must be repaired immediately.

## Signs of Leakage into a Low-Pressure System (e.g., Excessive Purging)

- Several signs can indicate a leak in a low-pressure system. These signs include:

1) Excessive purging of the purge unit.

2) High head pressure.

3) Moisture in the purge unit.

4) A rise in pressure during a standard vacuum test.

- If you notice any of the following signs, it is essential to inspect the system for leaks immediately.

## Maximum Leak Test Pressure for Low-Pressure Centrifugal Chillers

- The maximum test pressure for low-pressure centrifugal chillers is 10 psig. Pressures higher than this can damage the system.

- It is crucial to follow the instructions provided by the manufacturer when performing leak tests on low-pressure centrifugal chillers.

## Reporting for Chronically Leaking Appliances

- Chronic leaks in appliances must be reported to the EPA for repair.

- If you observe an appliance that is leaking continuously, you can report it to the Environmental Protection Agency (EPA) by dialing 1-800-424-8980.

# (B) LEAK REPAIR REQUIREMENTS

## Allowable Annual Leak Rate for Commercial and Industrial Process Refrigeration

The annual leak rate allowance for commercial and industrial process refrigeration is 35% of the annual charge. In other words, if an appliance has a 100-pound refrigerant charge, it can leak up to 35 pounds of refrigerant per year before needing repairs.

## Allowable Annual Leak Rate for Other Appliances Containing More Than 50 Pounds of Refrigerant

For appliances that contain more than 50 pounds of refrigerant, the allowable annual leak rate is 15% of the annual charge. If an appliance has 100 pounds of refrigerant, it can leak up to 15 pounds per year before it needs to be repaired.

## Other Important Information About Leak Repair Requirements

It is essential to fix any leaks within 30 days of discovering them. Failure to do so may require the owner or operator to retrofit or retire the appliance. The owner or operator is responsible for keeping records of all leak inspections and repairs for three years. It is important to repair leaks to protect the environment and workers from harmful refrigerant releases.

| Appliance Type | Allowable Annual Leak Rate |
|---|---|
| Commercial and Industrial Process Refrigeration | 35% of the charge per year |
| Other Appliances Containing More Than 50 Pounds of Refrigerant | 15% of the charge per year |

# (C) RECOVERY TECHNIQUES

## Recovering Liquid at the Beginning of the Recovery Process Speeds Up the Process

Recovering liquid refrigerant before it vaporizes can speed up recovery because it has a higher specific heat than vapor refrigerant, meaning it takes more energy to vaporize.

## Need to Recover Vapor in Addition to Liquid

It is important to recover both the liquid and vapor refrigerant from a system since vapor refrigerant can contribute to ozone depletion and global warming.

## Need to Heat Oil to 130°F Before Removing It to Minimize Refrigerant Release.

It is essential to fix any leaks within 30 days of discovering them. Failure to do so may require the owner or operator to retrofit or retire the appliance.

The owner or operator is responsible for keeping records of all leak inspections and repairs for three years.

It is important to repair leaks to protect the environment and workers from harmful refrigerant releases.

## Need to Circulate or Remove Water from Chiller During Refrigerant Evacuation to Prevent Freezing

During the refrigerant evacuation, it is crucial to either circulate or remove water from a chiller to avoid freezing.

This is because when the refrigerant is being eliminated from the system, it lowers the pressure inside the system, which can result in water freezing.

# High-Pressure Cut-Out Level of Recovery Devices Used with Low-Pressure Appliances

The high-pressure cut-out level for recovery devices used with low-pressure appliances is typically set at 10 psig to prevent damage to the system from high pressure.

Recovery techniques are crucial for safeguarding the environment and protecting workers' well-being. By following these techniques, you can prevent the release of refrigerants into the atmosphere and ensure worker safety.

| Technique | Description |
| --- | --- |
| Recover liquid refrigerant first | This will speed up the recovery process. |
| Recover both liquid and vapor refrigerant | This is important to prevent ozone depletion and global warming. |
| Heat oil to 130°F before removing it | This will help to release the refrigerant that is dissolved in the oil. |
| Circulate or remove water from chiller during refrigerant evacuation | This will prevent the water from freezing. |
| Set high-pressure cut-out level for recovery devices at 10 psig | This will prevent the pressure in the system from rising too high. |

## (D) RECHARGING TECHNIQUES

### Need to Introduce Vapor Before Liquid to Prevent Freezing of Water in the Tubes

Introducing vapor before liquid into a low-pressure refrigeration system is essential to prevent water from freezing in the tubes.

This is because liquid refrigerant has a much lower boiling point than vapor refrigerant. If liquid refrigerant is introduced into a vacuum, it will boil off quickly and may cause water in the tubes to freeze.

**Need to Charge Centrifugals Through Evaporator Charging Valve**

Centrifugal compressors are charged through the evaporator charging valve, which is the lowest point in the system and has the lowest refrigerant vapor pressure.

**Here are some additional tips for recharging low-pressure appliances:**

When charging the system with refrigerant, use a pressure/temperature chart and set it slowly to prevent boiling. Monitor closely for leaks and dispose of refrigerant properly.

Low-pressure appliances require proper recharging techniques to prevent damage and ensure worker safety.

**Here is a table that summarizes the key points of the recharging techniques for Type 3 (Low-Pressure) appliances:**

| Technique | Description |
| --- | --- |
| Introduce vapor before liquid | This will prevent the liquid refrigerant from freezing the water in the tubes. |
| Charge centrifugals through evaporator charging valve | This is the lowest point in the system and the point where the refrigerant vapor pressure is the lowest. |

# (E) RECOVERY REQUIREMENTS

## Evacuation Requirements for Low-Pressure Appliances in Disposal Situations

### Major Repairs

Any maintenance, service, or repair that removes any listed components is considered a major repair.

- Compressor
- Condenser
- Evaporator
- Auxiliary heat exchanger coil

### Non-Major Repairs

Any repair not involving the above-listed major components is a non-major repair.

### Leaky Appliances

If an appliance leaks refrigerant, it should be evacuated to the lowest possible level before major repair.

### Appliance or Component Containing Less Than 200 Pounds of Refrigerant

For appliances or components with less than 200 pounds of refrigerant, the required evacuation level is 0 inches Hg vacuum.

### Appliance or Component Containing More Than 200 Pounds of Refrigerant

Appliances or components with over 200 pounds of refrigerant must be evacuated to a 4-inch Hg vacuum.

# Recovery/Recycling Equipment Manufactured Before or After November 15, 1993

Equipment manufactured or imported before November 15, 1993, must be evacuated to 25 inches Hg, while equipment manufactured or imported on or after November 15, 1993, must be evacuated to 25 mm Hg absolute.

# Evacuation Requirements for Low-Pressure Appliances in Other Situations

## Major Repairs

The same evacuation requirements apply to low-pressure appliances as other types.

For **non-major repairs**, the allowable methods for pressurizing a low-pressure system are:

- Controlled hot water
- System heating/pressurization devices such as Prevac

The system pressure must be kept below 10 PSIG to avoid the release of refrigerant through the rupture disk, generally set at 15 psig.

## Need to Wait a Few Minutes after Reaching the Required Recovery Vacuum to See if System Pressure Rises

After achieving the necessary vacuum pressure, the technician should monitor the system pressure for a few minutes.

If the pressure increases, indicating the presence of refrigerant in the system, recovery must be repeated.

# (F) REFRIGERATION

## Purpose of purge unit in low-pressure systems

Low-pressure systems operate below atmospheric pressure, allowing air and moisture to easily enter the refrigerant circuit through leaks in gaskets, seals, or other components.

Non-condensable gases can create various issues, such as:

- Reduced system efficiency
- Increased compressor wear
- Damage to system components

A purge unit is a device that removes non-condensable gases from a low-pressure system. Typically, it consists of a vacuum pump, a condenser, and a receiver.

The vacuum pump eliminates air and moisture from the system, whereas the condenser condenses refrigerant vapor. The receiver stores the refrigerant until it is required.

The purge unit should be operated regularly to detect leaks. Excessive operation indicates leakage.

## Pressure-Temperature Relationships of Low-Pressure Refrigerants

- The pressure-temperature relationship of a refrigerant is the connection between pressure and temperature at a given state. Proper maintenance and operation of refrigeration systems rely on understanding this relationship.

- The pressure-temperature relationship of a refrigerant can be found on a pressure/temperature chart. This chart displays the pressure of a refrigerant at various temperatures.

- To utilize a pressure/temperature chart, locate the desired temperature on the chart. Once found, draw a straight, vertical line from the temperature until it intersects with the curve for the refrigerant you are using. The point at which the vertical line meets the curve indicates the pressure of the refrigerant at that temperature.

- To determine the pressure of refrigerant R-134a at a temperature of 40 degrees Fahrenheit, refer to the pressure-temperature chart. Locate the temperature of 40°F on the chart and draw a vertical line upwards until it intersects with the corresponding pressure value of 100 psig. This is the pressure of the refrigerant at 40°F.

- The pressure-temperature relationship of a refrigerant can be utilized to determine the refrigerant charge amount, troubleshoot system issues, and design new refrigeration systems.

| Situation | Appliance or Component Weight | Required Evacuation Level |
|---|---|---|
| Disposal of a low-pressure appliance | Any weight | 25 mm Hg absolute |
| Major repair of a low-pressure appliance | Any weight | 0 inches Hg vacuum or 25 mm Hg absolute |
| Non-major repair of a low-pressure appliance | Less than 200 pounds | 0 inches Hg vacuum |
| Non-major repair of a low-pressure appliance | More than 200 pounds | 4 inches Hg vacuum |
| Recovery/recycling equipment manufactured before November 15, 1993 | Any weight | 25 inches Hg |
| Recovery/recycling equipment manufactured on or after November 15, 1993 | Any weight | 25 mm Hg absolute |

- The concepts of the purge unit and the pressure-temperature relationship of low-pressure refrigerants are crucial for ensuring refrigeration systems' safe and efficient operation.

Placing the purge unit in a well-ventilated area is essential to prevent fume buildup. You should operate the purge unit at the correct pressure to prevent damage to the system. Regular inspections and maintenance are necessary to ensure the purge unit functions correctly.

## (G) SAFETY

### Equipment Room Requirements under ASHRAE Standard 15 (oxygen deprivation sensor with all refrigerants)

- According to the ASHRAE Standard 15, rooms containing refrigerants must have oxygen deprivation sensors and alarms. These sensors should be installed in the room to detect any drop in oxygen concentration below the TLV-TWA (Threshold Limit Value - Time Weighted Average). The alarms will sound when the oxygen concentration drops below the TLV-TWA, warning that the room needs to be evacuated immediately.

- Maintaining the oxygen concentration in a room above 19.5% is crucial, which is known as the TLV-TWA for oxygen. If the concentration falls below this level, individuals may suffer from symptoms like dizziness, headache, and nausea due to oxygen deprivation. In severe cases, oxygen deprivation could lead to unconsciousness and even death.

- The oxygen deprivation sensors and alarms should be tested regularly to ensure proper function and timely alerts for unsafe oxygen levels.

- ASHRAE Standard 15 mandates that equipment rooms housing R-123 should have a refrigerant sensor in addition to the oxygen deprivation sensor. The refrigerant sensor must be installed to detect any leak of R-123 and trigger an alarm once the refrigerant concentration reaches a certain level. At this point, the room should be immediately evacuated to ensure safety.

- According to the ASHRAE Standard 15, rooms containing refrigerants must have oxygen deprivation sensors and alarms. These sensors should be installed in the room to detect any drop in oxygen concentration below the TLV-TWA (Threshold Limit Value - Time Weighted Average). The alarms will sound when the oxygen concentration drops below the TLV-TWA, warning that the room needs to be evacuated immediately.

- Maintaining the oxygen concentration in a room above 19.5% is crucial, which is known as the TLV-TWA for oxygen. If the concentration falls below this level, individuals may suffer from symptoms like dizziness, headache, and nausea due to oxygen deprivation. In severe cases, oxygen deprivation could lead to unconsciousness and even death.

- The oxygen deprivation sensors and alarms should be tested regularly to ensure proper function and timely alerts for unsafe oxygen levels.

- ASHRAE Standard 15 mandates that equipment rooms housing R-123 should have a refrigerant sensor in addition to the oxygen deprivation sensor. The refrigerant sensor must be installed to detect any leak of R-123 and trigger an alarm once the refrigerant concentration reaches a certain level. At this point, the room should be immediately evacuated to ensure safety.

- R-123 is a refrigerant that does not catch fire, but it is categorized as a high-toxicity refrigerant. This implies that inhaling it in high concentrations can harm human health. Therefore, the refrigerant sensor plays a crucial role in detecting any leaks of R-123, enabling prompt evacuation of the room and timely repair of the leak.

To ensure the safety of equipment rooms containing refrigerants, it is important to follow these tips:

- Ensure proper ventilation in the room to prevent the buildup of fumes.
- Insulate the refrigerant lines properly to prevent condensation.
- Regularly inspect the refrigerant lines for leaks.
- Equip the room with fire extinguishers and other fire safety equipment.
- Provide training to the staff working in the room on safely handling refrigerants.

# 6. Practice Tests

# CORE SECTION

Q 1: Which of the below standards should be met before shipping any used refrigerant cylinders?

Answer :
(a) The Significant New Alternatives Policy (SNAP)
(b) Underwriters Laboratories
(c) Department of Transportation
(d) All of the above

Q 2: In case shipping used refrigerant cylinders, they must be:

Answer :
(a) Filled to no more than 80% of their capacity.
(b) Shipped upright.
(c) Shipped in a well-ventilated area.
(d) All of the above.

Q 3: Which international treaty regulates the production and use of ozone-depleting substances like CFCs, HCFCs, halons, methyl chloroform, and carbon tetrachloride?

Answer :
(a) Montreal Protocol
(b) Treaty of CFC
(c) Vienna Convention Law
(d) All of the above.

Q 4: Which of the below has a long-term effect due to the use and release of CFCs and HCFCs?

Answer :
(a) transportation
(b) heating equipment
(c) water supply
(d) the stratosphere

Q 5: Crop loss, eye diseases, and skin cancer have been linked to:

Answer :
(a) Depletion of Stratospheric oxygen
(b) Depletion of Stratospheric ozone
(c) Depletion of Stratospheric chlorine monoxide
(d) All of the above

Q 6: When chlorine reacts with ozone, it creates …………..and leaves behind an ………… molecule.

Answer :
(a) chlorine monoxide, oxygen
(b) carbon, chlorine
(c) carbon dioxide
(d) All of the above.

Q 7: This refrigerant has a GWP of "1" and has the same potential to contribute to climate change as carbon dioxide.

Answer :
(a) R-410A
(b) R-22
(c) R-1233zd
(d) Noneof the above.

Q 8: A refrigerant cylinder designed to hold recovered refrigerant has which color combination:

Answer :
(a) green body and blue top.
(b) Red body and yellow top.
(c) grey body and yellow top.
(d) green body and grey top.

Q 9: Which type of refrigerant can be transferred or sold to another individual?

Answer :
(a) Reclaimed
(b) Disposed
(c) Recovered
(d) All of the above

Q 10: The atom found in CFC and HCFC refrigerants that damages ozone in the stratosphere is ………………?

Answer :
(a) Hydrogen
(b) Chlorine
(c) Carbon
(d) Fluorine

Q 11: Which of the below refrigerant is a CFC?

Answer :
(a) R-410A
(b) R-12
(c) R-1233zd
(d) R-123

Q 12: After recovering refrigerant from a sealed system, what gas should be used to pressurize or blow debris out of the system for leak checking?

Answer :
(a) Hydrogen
(b) Oxygen
(c) Nitrogen
(d) None of the above

Q 13: What are the functions performed by oil in the compressor?

Answer :
(a) lubrication
(b) Seal the moving parts of a compressor
(c) Cooling
(d) A and B Both

Q 14: It is important to ensure that the refrigerant recovery cylinder is filled within................% of its capacity.

Answer :
(a) 40
(b) 80
(c) 70
(d) 55

Q 15: Refrigerant reusable containers must be hydrostatically tested, and date stamped every ............. years.

Answer :
(a) 5
(b) 2
(c) 6
(d) 3

Q 16: In a refrigerant system, the .......................... serves as the connection between the evaporator and the compressor.

Answer :
(a) Evaporation Line
(b) Suction Line
(c) Liquid Line
(d) None of the above

**Q 17: Why is it necessary to dehydrate a refrigeration system?**

**Answer :**
(a) To remove ice contaminates
(b) To remove water and water vapor contaminating the system
(c) To remove debris from the system
(d) Remove water from a hydrostatic pressure test

**Q 18: What are the consequences of leaving moisture in a refrigeration system still in operation?**

**Answer :**
(a) Create highly corrosive and toxic acids
(b) Freezes up
(c) Increases head pressure
(d) None of the above

**Q 19: What are the necessary steps before initiating an evacuation?**

**Answer :**
(a) Install dryer
(b) Recover all refrigerant
(c) Install site glass
(d) All of the above

**Q 20: What is the maximum number of refrigerants a recovery cylinder can store?**

**Answer :**
(a) One
(b) Three
(c) Four
(d) None of the above

Q 21: The service manifold's center port is typically identified by a ............... hose.

Answer :
(a) Yellow
(b) Blue
(c) Grey
(d) Green

Q 22: Which refrigerant is used in appliances with very high pressure?

Answer :
(a) R-500
(b) R-410A
(c) R-22
(d) R-503

Q 23: The chlorine in CFCs and HCFCs ................

Answer :
(a) will not dissolve in water.
(b) will rain out of the atmosphere.
(c) can be easily removed from the atmosphere.
(d) is not harmful to the environment.

Q 24: Which of the below is correct about the Significant New Alternatives Policy (SNAP) Program?

Answer :
(a) Covers only CFC and HCFC refrigerants used in industrial process refrigeration.
(b) Covers only CFC and HCFC refrigerants used in residential air conditioning.
(c) Relates to CFC and HCFC refrigerants used in commercial air conditioning.
(d) EPA's program that identifies refrigerants with lower overall risks to human health and the environment.

Q 25: In which phase will the refrigerant be in the discharge line?

Answer :
(a) Hot vapor
(b) Cold liquid
(c) Hot liquid
(d) Warm vapor


Q 26: What is the ASHRAE classification of slightly flammable substances?

Answer :
(a) B
(b) A3
(c) A2L
(d) A1


Q 27: Low-pressure systems have a pressure relief valve set at ......... psi.

Answer :
(a) 10
(b) 12
(c) 8
(d) 15

Q 28: ................... in the stratosphere breaks down CFCs, which releases chlorine ions.

Answer :
(a) X-radiation
(b) Gamma radiation
(c) Infrared radiation
(d) Ultraviolet radiation

**Q 29: How can you check a system for leaks?**

**Answer :**
(a) Use water pressure
(b) Pressurize with oxygen
(c) Use a hydrostatic test
(d) Evacuate the system and pull a vacuum on it

**Q 30: What steps should be taken if the oil has been contaminated due to burnout?**

**Answer :**
(a) Conduct a deep vacuum before recharging
(b) Install a permanent filter-drier
(c) Triple evacuate the system
(d) All the above

**Q 31: Could you explain the difference between recovery and recycling?**

**Answer :**
(a) There is no difference
(b) Recycle is to clean refrigerant for immediate reuse
(c) Recover is to remove refrigerant in any condition from a system and store it in an external container
(d) b&c

**Q 32: Which standard from below options should refrigerant recovery equipment meet?**

**Answer :**
(a) DOT standards and labeling
(b) CAA approval
(c) Certified and labeled by an EPA-approved equipment manufacturer
(d) All of the above

Q 33: Refrigeration technicians who violate the Clean Air Act may :

Answer :
(a) be fined
(b) lose certification
(c) appear in court
(d) All of the above


Q 34: After 1995, which option below is correct for obtaining CFC refrigerants to service existing refrigeration appliances in the US?

Answer :
(a) From solvent conversion.
(b) From European manufacturers.
(c) From recovery and recycling.
(d) All of the above

Q 35 : A reward of up to USD……………………… may be given for information that leads to a penalty against a technician who intentionally vents refrigerants into the atmosphere.

Answer :
(a) 10000
(b) 5000
(c) 8000
(d) 15000

Q 36: Which component changes low-pressure vapor to high-pressure vapor

Answer :
(a) compressor.
(b) evaporator.
(c) condenser.
(d) None

Q 37: Usually, blended refrigerants should be charged ............

Answer :
(a) at very high temperatures
(b) as liquid
(c) as vapor
(d) None of the above

Q 38: Manifold gauge sets are color-coded with the high-pressure gauge in ............

Answer :
(a) yellow
(b) grey
(c) blue
(d) red

Q 39 : When disposing of a disposable refrigerant cylinder, ensure that the internal pressure is reduced to ...... psig.

Answer :
(a) 0
(b) 5
(c) 8
(d) 15

Q 40: Refrigerants cannot be called "reclaimed" unless they

Answer :
(a) are recovered using equipment less than 1 year old.
(b) were properly processed in a reclaiming machine.
(c) have been chemically analyzed, and the refrigerants meet the ARI-700 purity standard.
(d) None

Q 41: What are some of the most severe consequences of damaging the Earth's protective ozone layer?

Answer :
(a) An increase in natural radioactivity.
(b) Marine plant mutations.
(c) An increase in volcanic activity.
(d) The incidence of skin cancer in humans is on the rise.

Q 42: Which is used to remove ice from a sight glass?

Answer :
(a) R-11 refrigerant
(b) water
(c) An alcohol spray
(d) All of the above

Q 43: Recovery equipment that relies on the refrigerant pressure within the appliance is considered................

Answer :
(a) illegal.
(b) system-dependent.
(c) self-sufficient.
(d) self-contained.

Q 44: The most commonly used oil in stationary air conditioning systems with HFC-410a refrigerant is ..............
Answer :
(a) ester oil.
(b) mineral oil.
(c) alkylbenzene oil.
(d) None

Q 45: What do chlorofluorocarbon (CFC) and hydrofluorocarbon (HCFC) refrigerants have in common?

Answer :
(a) They have the same normal boiling points.
(b) It is important to recover both of them before opening appliances for servicing or disposing of them.
(c) They both contain hydrogen.
(d) They have identical ODP.

Q 46: The refrigeration system's compressor receives a ………………… as its input refrigerant.

Answer :
(a) superheated vapor.
(b) subcooled vapor.
(c) subcooled liquid
(d) liquid.

Q 47: What steps is the U.S. taking to protect the stratospheric ozone layer?

Answer :
(a) Capturing and ultimately eliminating the use of chlorofluorocarbons.
(b) Enforcing strict emission requirements on garbage incinerators.
(c) Replacing coal-fired power plants for the production of electricity.
(d) All of the above.

Q 48: The process of a refrigerant blend boiling or condensing over a range of temperatures at constant pressure is known as ……………..

Answer :
(a) temperature glide.
(b) mixture slump.
(c) pressure slide.
(d) All of the above.

**Q 49: Which refrigerant contains a CFC?**

Answer :
(a) R-404A
(b) R-152a
(c) R-410a
(d) R-12


**Q 50: Which of the below is illegal?**

Answer :
(a) manufacture or import HFC refrigerants into the U.S.
(b) use CFC refrigerants.
(c) releasing any refrigerant during the service maintenance repair. or disposal of appliances except for de minimis releases.
(d) Any of the above

# ANSWERS

## CORE SECTION

1. d
2. d
3. a
4. d
5. b
6. a
7. c
8. c
9. a
10. b
11. b
12. c
13. d
14. b
15. a
16. b
17. b
18. a
19. b
20. a
21. a
22. d
23. a
24. d
25. a
26. c
27. d
28. d
29. d
30. d
31. d
32. c
33. d
34. c
35. a
36. a
37. b
38. d
39. a
40. c
41. d
42. c
43. b
44. a
45. b
46. a
47. d
48. a
49. d
50. c

# TYPE 1

**Q 1: Could you clarify the EPA's definition of a small appliance?**

Answer :
(a) Manufactured, charged, and hermetically sealed in a factory and contains five pounds or less of refrigerant
(b) Manufactured, charged, and hermetically sealed in a factory and contains fifteen pounds or less of refrigerant
(c) Manufactured, charged, and hermetically sealed in a factory and contains two pounds or less of refrigerant
(d) None of the above

**Q 2: What is the most commonly used refrigerant to replace R-12 in new household refrigerators?**

Answer :
(a) R-141a / 141b blend
(b) R-134a
(c) R-123
(d) None of the above.

**Q 3: Recovery equipment used for servicing or repairing small appliances must be certified by an EPA-approved laboratory if the equipment was manufactured after ...........**

Answer :
(a) May 15, 1993
(b) June 15, 1992
(c) November 15, 1993
(d) July 15, 1993

**Q 4: What is the maximum amount of refrigerant allowed to be charged in Type I appliances at the factory?**

Answer :
(a) 2 pounds
(b) 3 pounds
(c) 4 pounds
(d) 5 pounds

**Q 5: What are some possible causes of high pressure on the discharge side of an active self-contained recovery unit?**

Answer :
(a) A closed recovery tank outlet valve.
(b) A closed recovery tank inlet valve.
(c) Excessive air or other non-condensable in the recovery tank.
(d) B & C

**Q 6: What refrigerants must be recovered with EPA-regulated equipment under section 608?**

Answer :
(a) R-12
(b) R-134a
(c) R-123
(d) Sulfur Dioxide

**Q 7: In the event of a significant release of R-12 or R-22 in a contained area, what should you do?**

Answer :
(a) Respiratory protection is not required as these gases are not toxic.
(b) Dust masks are sufficient protection against harmful vapors.
(c) Self-contained breathing apparatus (SCBA) or leaving the area is required.
(d) All of the above

**Q 8: The EPA requires appliances to have a service aperture when adding or removing refrigerant. For small appliances, this service port is typically................**

Answer :
(a) located 15 inches below the compressor.
(b) installed at the factory and is 1/4 inch in diameter with machine threads.
(c) A piercing access valve can be used to enter the system through a straight piece of tubing known as a Process Tube.
(d) None of these

**Q 9: When is it appropriate to use piercing valves, and what is a common issue with them?**

Answer :
(a) Piercing valves are not allowed.
(b) A closed recovery tank outlet valve.
(c) Fittings tend to leak over time and should not be left on appliances
(d) B & C

**Q 10: CFCs will no longer be produced within the borders of the United States after:**

Answer :
(a) 1996
(b) 1995
(c) 1997
(d) 2000

**Q 11: After November 15, 1993, recovery equipment for small appliances must be capable of recovering:**

Answer :
(a) 99% of the refrigerant regardless of compressor operation and achieve a 10-inch vacuum under ARI 740-1993
(b) 80% of the refrigerant when the compressor is not running or achieves a 4-inch vacuum under ARI 740-1993.
(c) 90% of refrigerant when the compressor is operating or achieves a 4-inch vacuum under ARI 740-1993.
(d) b & c

**Q 12: Passive refrigerant recovery of small appliances, dependent on the system.**

Answer :
(a) It is important to recover refrigerant in a container that is not pressurized.
(b) It is necessary to recover 80% of the refrigerant.
(c) It is not necessary to have a functioning compressor.
(d) All of the above

Q 13: Should the Clean Air Act (CAA) regulations be modified after a technician obtains certification?

Answer :
(a) It will be the technician's responsibility to learn and comply with future changes in the law.
(b) The technicians must be retested and pass the exam with 84%.
(c) The technician must take a new test to be recertified. .
(d) B & C

Q 14: If you have a small appliance that contains R-12, it is important to take proper steps before disposing of it.

Answer :
(a) Turn upside down
(b) Pressurize with nitrogen
(c) Recover the refrigerant
(d) All of these

Q 15: When servicing small appliances, individuals recovering refrigerant must hold certification as a Universal Technician or a ................

Answer :
(a) Type II Technician
(b) Core Technician
(c) Type I Technician
(d) None of the above

Q 16: To work on small appliances manufactured after November 14, 1993, a technician must hold which certification?

Answer :
(a) Universal
(b) Type 1
(c) Type 2
(d) a & b

Q 17: What is the ideal temperature to check for excess air in a cylinder?

Answer :
(a) 52F
(b) 55F
(c) Room temperature before taking a pressure reading.
(d) b & c

Q 18: EPA regulations mandate the recovery of at least 80% of refrigerant from small appliances with non-operating compressors if the technician is using ...........

Answer :
(a) a system-dependent (passive) recovery unit.
(b) a self-contained (active) recovery unit.
(c) either type of recovery unit.
(d) None of these

Q 19: What should you do when removing refrigerant from a frost-free refrigerator?

Answer :
(a) turn on the defrost heater to vaporize any trapped liquid.
(b) heat the recovery cylinder to vaporize liquid refrigerant.
(c) pack ice around the evaporator to ensure maximum liquid is available.
(d) All of the above

Q 20: If exposed to high temperatures from open flames or hot metal surfaces, R-12 or R-22 can decompose into ................

Answer :
(a) sulfuric and phosphoric acids.
(b) boric and chromic acids.
(c) hydrochloric and hydrofluoric acids.
(d) None

**Q 21:** When using a graduated charging cylinder, any refrigerant that is vented ..................

**Answer :**
(a) does not need to be recovered.
(b) is considered a de minimis release.
(c) must be recovered.
(d) None of the above.

**Q 22:** How often must renew the EPA section 608 certification?

**Answer :**
(a) in 3 years
(b) in 5 years
(c) Doesn't expire
(d) 4 years

**Q 23:** Before beginning the recovery procedure, do not install and open a piercing access valve if the system is at ........ psig.

**Answer :**
(a) 5
(b) 4
(c) 1
(d) 0

**Q 24:** It is an EPA regulation that anyone who opens an appliance for maintenance, service, or repair must have at least one self-contained recovery machine available at their place of business. The only exception to this rule are persons working on :

**Answer :**
(a) high -pressure appliances.
(b) R-410a systems.
(c) small appliances.
(d) None

Q 25: Piercing-type valves should only be used on which of the following types of tubing materials?

Answer :
(a) Iron
(b) Copper
(c) Aluminum
(d) b&c

# ANSWERS
## TYPE 1

1. a
2. b
3. c
4. d
5. d
6. a
7. c
8. c
9. c
10. b
11. d
12. d
13. a
14. c
15. c
16. d
17. c
18. c
19. a
20. c
21. c
22. c
23. d
24. c
25. d

# TYPE 2

**Q 1:** To ensure proper operation, heat pumps that use R-410A refrigerant must undergo a leak check using .............

**Answer :**
(a) compressed oxygen.
(b) HFCs
(c) pressurized nitrogen.
(d) None of the above

**Q 2:** Dry nitrogen is recommended for breaking the initial vacuum during double evacuation dehydration. However, it ................

**Answer :**
(a) can be dangerous if not used with a pressure
(b) is expensive.
(c) is toxic.
(d) None of the above.

**Q 3:** A ................ should protect every air conditioning or refrigeration system.

**Answer :**
(a) low-pressure control.
(b) properly located stop valve.
(c) pressure relief device.
(d) None of these

**Q 4: Pressure relief valves must not be ...........**

**Answer :**
(a) installed vertically.
(b) Installed in series.
(c) installed in parallel.
(d) All of these

Q 5: To handle R-410a refrigerant, it is necessary to use recovery and recycling equipment that is certified and rated for a minimum of ........ psig.

Answer :
(a) 350
(b) 400
(c) 200
(d) None of the above

Q 6: Soap bubble testing is a technique used to ...............

Answer :
(a) to detect water leaks.
(b) only with CFCs and HCFCs.
(c) detect refrigerant leaks accurately.
(d) None of the above.

Q 7: When a refrigerant leaves the receiver of a system, it does so in the form of a ...............

Answer :
(a) Superheated liquid
(b) high pressure liquid.
(c) Subcooled vapor
(d) low pressure liquid.

Q 8: A high-pressure appliance uses a refrigerant with a liquid phase saturation pressure above 355 psia at 104F. This definition includes but is not limited to appliances using which of the refrigerants listed below?

Answer :
(a) R-503
(b) R-23
(c) R-13
(d) All of the above

Q 9: A maintenance task for refrigerant recycling machines is ..............

Answer :
(a) changing the filter and compressor oil, if needed.
(b) check compressor seals.
(c) change electrical fuses.
(d) None of the above

Q 10: It is necessary to remove refrigerant from the condenser outlet if the ........................

Answer :
(a) condenser is on the roof.
(b) condenser is below the receiver.
(c) compressor is inoperative.
(d) None of the above.

Q 11: When evacuating a system, using a large vacuum pump could ...............

Answer :
(a) cause trapped water to freeze.
(b) causes the valves to freeze.
(c) cause trapped refrigerant to freeze
(d) None of the above.

Q 12: The most common water source used for cooling a water-cooled recovery unit's condenser is the....................

Answer :
(a) de-ionized water.
(b) chilled water.
(c) local municipal water supply.
(d) All of the above

Q 13 : The required evacuation level for recovery equipment manufactured after November 15, 1993, on a system containing less than 200 pounds of R-12 refrigerant is ..............

Answer :
(a) 2 inches Hg
(b) 4 inches Hg
(c) 14 inches Hg
(d) 10 inches Hg

Q 14: What is the function of a filter drier in an HVAC system and when is it recommended to replace it?

Answer :
(a) To prevent moisture buildup in the refrigerant, replace the filter regularly and whenever the system is opened.
(b) Dryer's air filters. Replace yearly
(c) To filter the oil. Change every year.
(d) None of the above.

Q 15: What is the purpose of a filter drier and when is it necessary to replace it?

Answer :
(a) Dryer's air filters. Replace yearly
(b) To filter the oil. Change every year.
(c) To prevent moisture buildup in the refrigerant, replace the filter regularly and after any system maintenance.
(d) None of the above.

Q 16: Industrial and commercial refrigeration equipment over 50 pounds with an annual leak rate of ............% requires repair under EPA regulations.

Answer :
(a) 5
(b) 20
(c) 30
(d) 35

Q 17 : Before charging a water-cooled R-410a system, charge with vapor to at least .............. before switching to liquid charging.

Answer :
(a) 20 psig
(b) 120 psig
(c) 55 psig
(d) 40 psig

Q 18: Recovery tanks used for R-410a must have a minimum rating of ................... psig.

Answer :
(a) 400
(b) 300
(c) 200
(d) None of the above.

Q 19: What is the minimum vacuum pressure in inches of Mercury required for HCFC-22 in appliances with over 200lbs of refrigerant-using equipment before 11/15/1993?

Answer :
(a) 8
(b) 4
(c) 12
(d) None of the above.

Q 20: When evacuating a vapor compression system, what vacuum level is required in microns?

Answer :
(a) 50
(b) 200
(c) 500
(d) 350

Q 21: What does 'inches HG' mean

Answer :
(a) Inches on mercury
(b) Inches on hygrometer
(c) Inches on high gauge
(d) None of these

Q 22: If a unit contains more than ............. pounds of refrigerant charge, the owner is responsible for maintaining records of all refrigerant added.

Answer :
(a) 30
(b) 50
(c) 20
(d) None of the above.

Q 23: The majority of the liquid to be recovered will be found in the ...............

Answer :
(a) Evaporator
(b) Receiver (when applied)
(c) Condenser
(d) None of the above.

Q 24: Under EPA regulations, industrial and commercial refrigeration equipment with over 50 pounds of refrigerant and an annual leak rate of ........% must be repaired.

Answer :
(a) 20
(b) 20
(c) 50
(d) 35

**Q 25: Whenever a unit is being serviced, it is necessary to replace the ........**

**Answer :**
(a) filter-drier
(b) metering device
(c) thermostat
(d) b&c

# ANSWERS
## TYPE 2

1. c
2. a
3. c
4. b
5. b
6. c
7. b
8. d
9. a
10. b
11. a
12. c
13. d
14. a
15. c
16. d
17. b
18. a
19. b
20. c
21. a
22. b
23. b
24. d
25. a

# TYPE 3

Q 1: Where are the most common locations for leaks to occur in low-pressure systems?

(a) Gaskets or fittings
(b) Purge unit
(c) Evaporator
(d) None of the above

Q 2: Where are the common areas for leaks in open-drive compressor systems?

Answer :
(a) Cooling tower
(b) Shaft seal
(c) Condenser
(d) None of the above.

Q 3: When is it recommended to perform a leak check on a low-pressure system as per the ASHRAE guideline 3-1996?

Answer :
(a) low-pressure control. During pressure checks
(b) pressure relief device.
(c) Both a & b
(d) During vacuum testing, if the pressure rises above 2.5 mm Hg from its initial reading of 1 mm Hg

Q 4: What methods can be employed to raise the pressure within the system?

Answer :
(a) Increasing the flow rate of the fluid through the system
(b) controlled hot water
(c) heater blankets
(d) Both b & c

Q 5: What are the potential consequences if the system is pressurized beyond 10 psig?

Answer :
(a) Oxygen combined with oil in the system will explode
(b) Shaft seals will fail
(c) The ruptured disc will fail
(d) None of the above

Q 6: What methods or techniques can be used to raise the pressure in the system?

Answer :
(a) heater blankets
(b)  controlled hot water
(c)  Both a & b
(d) None of the above.

Q 7: At what temperature would Refrigerant-11 (R-11), at 14.7 psia, will boil?

Answer :
(a) 84.5 F
(b)  74.5 F
(c)  64.5 F
(d) None of these

Q 8: As per ASHRAE Standard 15-1994, which safety group classification mandates the installation of refrigerant sensors in equipment rooms?

Answer :
(a) A2
(b) A1
(c) B2
(d) All refrigerant safety groups.

Q 9: A hydrostatic tube test kit is a tool used to ……………………

Answer :
(a) determine if a tube leaks.
(b) blow all water from the tubes.
(c) remove any water from the machine.
(d) None of the above

Q 10: When evacuating a system, using an excessively large vacuum pump can cause …………………..

Answer :
(a) trapped oil to freeze.
(b) valves to freeze.
(c) trapped water to freeze.
(d) None of the above.

Q 11: What measures can be taken to prevent the accumulation of air in a low-pressure refrigeration system that is not in use?

Answer :
(a) Intermittently operate the system with no load.
(b) Leave the purge unit on at all times.
(c) Maintaining the system pressure slightly above the atmospheric pressure is important.
(d) None of these

Q 12: When the leak rate exceeds ……………percent of the charge per year, commercial and industrial process refrigeration systems must be repaired.

Answer :
(a) 30
(b) 20
(c) 10
(d) None of the above

Q 13: According to the EPA, an average 350-ton R-11 chiller at 0 psig pressure still contains about …………of refrigerant vapor after the liquid R-11 has been removed.

Answer :
(a) 30 lbs
(b) 20 lbs
(c) 100 lbs
(d) None of the above

Q 14: To determine how much vapor to add to a vacuum chiller, charge with vapor until...

Answer :
(a) The recovery unit pressure drops.
(b) valves to freeze.
(c) the system pressure corresponds to a refrigerant saturation temperature of 36 F.
(d) None of the above.

Q 15: The discharge from a ruptured disc should be piped outdoors for venting

Answer :
(a) outdoors
(b) to the duct system
(c) to the evaporator
(d) None of these

Q 16: A typical setting for the high-pressure cut-out control on a recovery unit used to evacuate refrigerant from a low-pressure chiller is typically set at ………

Answer :
(a) 10 psig
(b) 20 psig
(c) 15 psig
(d) None of the above

Q 17: To efficiently check for leaks in a low-pressure refrigeration system, it can be charged first. Then, .................. can be used for the purpose.

Answer :
(a) raising system pressure by heating with circulated hot water or heating blankets.
(b) adding HCFC-22.
(c) operating the purge system.
(d) None of the above

Q 18: As per the EPA regulations for a non-major repair,........................ is used to pressurize a low-pressure chiller to open the system.

Answer :
(a) )warming the refrigerant
(b) adding compressed air
(c) adding carbon dioxide
(d) None of the above.

Q 19: After all of the R-11 liquid has been removed, an average 350-ton R-11 chiller at 0 psig pressure still contains approximately ................. of refrigerant vapor, according to the EPA.

Answer :
(a) 50 lbs
(b) 60 lbs
(c) 100 lbs
(d) None of these

Q 20: The purge unit takes its suction from the.............................

Answer :
(a) evaporator outlet.
(b) compressor discharge.
(c) top of the condenser.
(d) None of the above

Q 21: Where does the refrigerant go after leaving the purge unit of a low-pressure centrifugal system?

Answer:
(a) 30 lbs
(b) 20 lbs
(c) 100 lbs
(d) None of the above

Q 22: Could you clarify which repairs are always classified as "major" repairs according to the EPA's regulations?

Answer:
(a) Replacement of a filter-drier.
(b) Replacement of a purge unit.
(c) Replacement of an evaporator coil.
(d) None of the above.

Q 23: Commercial and industrial process refrigeration systems must be repaired when the leak rate exceeds …….percent of the annual charge.

Answer:
(a) 30
(b) 20
(c) 35
(d) None of these

Q 24: What is the lowest access point on the low-pressure centrifugal air-conditioning unit?

Answer:
(a) The evaporator charging valve
(b) The purge unit exhaust valve
(c) The condenser service valve
(d) None of the above

**Q 25:** The hydrostatic test checks ………………….. within a chiller.

**Answer :**
(a) Tubing leaks
(b) Moisture leaks
(c) a&b
(d) Compressor failure

# ANSWERS
## TYPE 3

1. a
2. b
3. d
4. d
5. c
6. c
7. b
8. d
9. a
10. c
11. c
12. a
13. c
14. c
15. a
16. a
17. a
18. a
19. c
20. c
21. c
22. c
23. a
24. a
25. a

# GLOSSARY

- **Refrigerant:** A refrigerant is a substance used in a refrigeration system to absorb and release heat.

- **Ozone Depletion Potential (ODP):** The term "ozone depletion potential" refers to how much a refrigerant can harm the ozone layer. Let me know if you need any further assistance.

- **Global Warming Potential (GWP):** The Global Warming Potential (GWP) is a metric used to measure the impact of a refrigerant on global warming in comparison to carbon dioxide.

- **Stratospheric Ozone:** The ozone layer in the Earth's stratosphere protects against harmful ultraviolet (UV) radiation.

- **Recovery:** The process of removing refrigerant from a system for reuse or recycling.

- **Recycling:** The process of cleaning and reprocessing extracted refrigerants to reuse them in the same system.

- **Reclamation:** The process of purifying refrigerants to meet the industry standards for resale involves removing any impurities and contaminants that may have been present in the refrigerant. This ensures that the refrigerant is safe to use and meets the required specifications for its intended use.

- **EPA 608 Certification:** A certification is required for individuals who handle refrigerants to ensure compliance with environmental regulations.

- **Leak Rate:** The measurement of refrigerant leakage is expressed as pounds per year (lbs/yr).

- **Pressure-Enthalpy Chart:** A graphical representation used to analyze refrigerant properties in a system.

- **High-Pressure Refrigerants:** Residential air conditioning units usually use refrigerants that operate at pressures above 300 psig.

- **Low-Pressure Refrigerants**: Refrigerants are used in cooling systems operating at pressures below 300 psig, commonly found in chillers.

- **Appliance**: A device that utilizes refrigerants, such as air conditioners and refrigerators.

- **Leak Detection**: Identifying and locating refrigerant leaks in a system.

- **Refrigerant Recovery Cylinder**: A refrigerant recovery cylinder is a container for storing and transporting recovered refrigerants.

- **Dehydration:** The removal of moisture from a refrigeration system.

- **Medium-Pressure Refrigerants**: Refrigerants commonly used in commercial and industrial applications operate at medium pressures.

- **Superheat**: The temperature at which a refrigerant vapor turns into a liquid state after boiling.

- **Subcooling:** The temperature of the refrigerant liquid is lower than its condensing temperature.

- **Recovery Unit**: A refrigerant recovery machine extracts refrigerant from the system during service.

- **Thermal Expansion Valve (TXV)**: A component that controls refrigerant flow into the evaporator.

- **Low-Pressure Chillers**: Low-pressure refrigerants are commonly used for large cooling applications in chiller systems.

- **Pressure Relief Device**: A pressure relief device that is intended to discharge any excess pressure from a system.

- **Refrigerant Migration:** Refrigerant moves from high-pressure areas within a system to areas of lower pressure, which creates a cooling effect.

- **Evacuation**: The process of removing air and moisture from a refrigeration system before charging it with refrigerant is known as evacuation.

- **Charging**: Add refrigerant to a system to reach the desired pressure and temperature.

- **Condenser Water Loop:** A system that circulates water to remove heat from the condenser. There were no spelling, grammar, or punctuation errors to correct in the original text.

- **Compressor**: The compressor is a vital component of a refrigeration system. Its function is to compress the refrigerant gas, which is essential for the cooling process to occur.

# CONCLUSION

Congratulations on finishing the book!

As you finish this book and become a certified expert in HVAC systems, it's important to reflect on your journey. You've gained knowledge, skills, and a commitment to environmental responsibility.

Throughout your studies, you've learned about the impact of refrigerants on the environment, including ozone depletion potential and global warming potential. This book has been a faithful guide, providing you with the resources to become an expert in the field.

You've learned about responsible refrigerant management, including compliance and sustainability. Your knowledge of various refrigeration systems, equipment, and safety practices makes you a well-rounded professional in the field.

You've mastered Type I, II, III, and Core refrigeration systems with expertise in high/low-pressure and compressors/heat exchangers, making you a well-rounded professional in the field.

Good luck on your EPA 608 certification exam, and remember that you are part of a noble mission to make the world cooler and safer, one refrigerant molecule at a time.

*We'd Love Your Feedback!*

Please let us know how we're doing by leaving us a review.

www.ingramcontent.com/pod-product-compliance
Lightning Source LLC
Chambersburg PA
CBHW082209070526
44585CB00020B/2341